LIVING BRIDGES

Royal Academy of Arts, London

26 September – 18 December 1996

LIVING BRIDGES

The inhabited bridge, past, present and future

Edited by
Peter Murray and Mary Anne Stevens
With contributions by
David Cadman, Jean Dethier, Ruth Eaton,
Stuart Lipton and Peter Murray

Royal Academy of Arts, London

Prestel Munich · New York

Administration

This exhibition has been developed from a concept initiated by, and has drawn upon research undertaken by the Centre Georges Pompidou, Musée national d'art moderne – Centre de création industrielle, Paris, under the direction of Jean Dethier.

It has been expanded by the Royal Academy of Arts under the guidance of Peter Murray and MaryAnne Stevens.

Exhibition

Curators
Peter Murray and MaryAnne Stevens

Advisor
Jean Dethier

Historical Curator
Ruth Eaton

Coordinators
Esther Waterfield with Emeline Max

Designer
Branson Coates

Project Designer
Tyeth Gundry

Tank Advisor
Tom Chapman (GU Projects)

Model Makers
Andrew Ingham and Associates

Catalogue Editors
Peter Murray and MaryAnne Stevens with Simon Haviland

Catalogue Coordinators
Sophie Lawrence with Katharine Klitgaard

Photographic Coordinators
*Miranda Bennion with Patricia Eaton, Sam Oakley,
Amanda Simmons, Roberta Stansfield and Esther Waterfield*

Thames Water Habitable Bridge Competition

Administrators
*Peter Murray
with Esther Waterfield and Katharine Klitgaard*

Assessors
Sir Philip Dowson, CBE PRA, Chairman
The Rt. Hon. John Gummer, MP,
Secretary of State for the Environment
Michael Cassidy, Corporation of London
Sir Robert Clarke, Thames Water Plc
Jean Dethier, Centre Georges Pompidou
Gordon Graham, CBE, MA, PPRIBA
Peter Murray
Janet Street-Porter

Acknowledgements

Peter Murray and MaryAnne Stevens wish to thank all those who have helped to realize the exhibition, and the architectural competition and the book which accompanies them. Apart from the people listed below, they would like to express their gratitude to Norman Rosenthal for his contribution to the design concept of the exhibition, to Jill Lever for her research on many of the bridges, to Liam O'Connor for his constant support of both the exhibition and the competition, and to the experts who provided invaluable input to the competition brief: Eddie Oliver, Jonathan Percival and John Taylor of KPMG; John Roberts and Andrew Ogden of Allott and Lomax, and Rear-Admiral Bruce Richardson, Fraser Clift and Gordon Dickins of the Port of London Authority. Special thanks are also due to Michael Maegraith, Simon Williams, Christopher Wynne, and Simon Haviland who have patiently guided the accompanying book through to publication.

Jean Dethier and Ruth Eaton would particularly like to thank Vittorio Gregotti and Dario Matteoni who gave them the opportunity to edit the special issue of *Rassegna* published in December 1991 (no. 48). They would also like to thank all those who contributed either to the issue of *Rassegna* or to earlier or later stages of research between 1990 and 1996. Finally they, too, express sincere gratitude to Liam O'Connor.

Jean Paul Baeten
Clare Both
Remy Bouguennec
M Boudry
John Brownjohn
Marasin Cavagna
Isabelle Chemouny-Smadja
Stewart Drew
Ellen Dunham Jones
Robert Elwall
Nicolas Faucherre
Paul Finch
Gilbert Gardes
Jean-Philippe Garric
Elisabetta Gonzo
Janette Harris
John Harris
Rosemary Harris

Catherine Howe
Andrew Ingham
Annie Jacques
Betrand Lemoine
Maud le Thery
Jill Lever
Jean Manco
Bernard Marrey
Jean Mesqui
Gilles Michon
Miron Mislin
Paolo Marochiello
Gavin Morgan
Valerie Negre
Andrew Norris
Sue Palmer
Philippe Panerai
Antoine Picon

Virginia Picon-Lefebvre
Jean-Claude Planchet
Paolo Polledri
Philippe Potié
Wellington Reiter
Patrick Renaud
Margaret Richardson
Guiseppina Carla Romby
Ionel Schein
Jeremy Smith
Henri Stierlin
Daniel Treiber
Alice Uebe
Jocelyne Van Deputte
Alessandra Vicari
Thomas Wellman
Carol Willis

First published on the occasion of the exhibition
'Living Bridges: The inhabited bridge, past, present and future'
Royal Academy of Arts, London, 26 September–18 December 1996

Exhibition realized with the collaboration of the
Centre Georges Pompidou, Musée national d'art moderne
– Centre de création industrielle, Paris

Exhibition supported by and

in association with **THE INDEPENDENT**

Additional support has been received from
Hammerson UK Properties plc, owners and developers of
Globe House
London Borough of Southwark, as part of its commitment
to the Cross River Partnership
The Drue Heinz Trust

The exhibition incorporates the

 Thames Water Habitable Bridge Competition

Front cover: *London Bridge,* engraving 1735; Nicolas Raguenet and
Joan Baptiste, *La joute des mariniers entre le Pont Notre-Dame et le
Pont au Change en 1756*, Musée Carnavalet, Paris; Zaha Hadid,
Model, © Collection Zaha Hadid; Antoine Grumbach, *The Garden
Bridge*, 1996. © Antoine Grumbach. Back cover: Bernard Tschumi,
Lausanne Bridge – city, project drawing, 1988, © Bernard Tschumi

Frontispiece: Canaletto, *Ponte di Rialto* (detail, see fig. 91, p. 71),
Musée du Louvre, Paris

Jean Dethier's essay translated from the French by John Brownjohn
Copy-edited by Simon Haviland

Catalogue prepared for Prestel by Simon Williams
Designed and typeset by Tom Harwerth and Kristiane Klas
Frankfurt am Main, Germany
Lithography by Fischer Repro, Frankfurt
Printed by Peradruck Matthias GmbH, Gräfelfing
Bound by R. Oldenbourg GmbH, Munich

ISBN: 3-7913-1734-2 (trade edition)

Printed in Germany
Printed on acid-free paper

Contents

Foreword by the President of the Royal Academy

Living Bridges *explores a fascinating building type, the inhabited bridge. This can be defined broadly as a bridge which not only provides a link between two points for pedestrian and vehicular traffic but also supports superstructures that can serve residential, commercial, religious, industrial or defensive purposes, thereby creating a continuity of the built-up area from one river bank to the other.*

From the twelfth century to the end of the eighteenth there were many inhabited bridges in Europe, including such celebrated examples as the Ponte Vecchio in Florence, the Ponte di Rialto in Venice, Old London Bridge and the bridges which once linked Paris' Île de la Cité to both Left and Right Banks.

The exhibition celebrates the history of inhabited bridges: scale models set across a 'river of time' demonstrate the evolution of the inhabited bridge from Old London Bridge to proposals to span the Seine, the Thames and the Arno in the twentieth century. The exhibition also addresses the viability of building a new inhabited bridge as part of the current debate surrounding the regeneration of the River Thames through London. To this end, we have held a limited international architectural competition for an inhabited bridge to span the river on a site between Waterloo Bridge and Blackfriars Bridge. The winning models of the competition provide the climax to the exhibition.

From its inception the exhibition has enjoyed the support of the Secretary of State for the Environment and Minister for London, the Rt. Hon. John Gummer MP. We are enormously indebted to him and to his staff, especially Liam O'Connor, for their assistance in arranging a feasibility study for the competition to be undertaken by KPMG. The Thames Water Habitable Bridge competition was made possible through the support of Thames Water, who provided most generous financial backing to enable models of the joint winners to be shown. The exhibition itself has been made possible through the generosity of the Corporation of London and Générale des Eaux, with media support being provided by The Independent. *We are most grateful to these organisations for their enthusiastic support of so complex an exhibition.*

Finally, this project is the first major collaboration between the Royal Academy of Arts and the Centre Georges Pompidou (MNAM-CCI). Throughout its planning we have worked with Jean Dethier and his assistant Ruth Eaton. Peter Murray and MaryAnne Stevens acted as co-curators of the exhibition for the Royal Academy; Nigel Coates provided the design of the river running through the galleries. We are most grateful to them all.

SIR PHILIP DOWSON CBE

9

Foreword by the Secretary of State for the Environment and Minister for London

We have not forgotten Old London Bridge history has seen to that and folklore to its survival. We are reminded time and time again by the tour guides of its importance in London's history; its stone street flanked by houses of differing styles and ages; the ancient chapel of St. Thomas of Canterbury and the great stone gateway. We know too that all the river crossings in London today are dominated by motor traffic, and that people endure a noisy and unfriendly walk to the middle of London's bridges to glimpse the magnificent views up and down the river.

Why should Londoners and visitors alike not have a bridge to themselves? A street over the river on which to promenade, free of traffic, with shops, cafés and other uses. Such a bridge, if built, should be a destination in its own right, not just infrastructure but a civic place on the River Thames drawing people from both banks like the Charles Bridge in Prague, the Ponte Vecchio in Florence or the Ponte di Rialto in Venice.

Opportunities to work with other European colleagues are always exciting. This has been especially true working with the Centre Georges Pompidou in Paris over the last two years on this project which is so relevant to London.

I was also very encouraged to find, early on, such willing and enthusiastic sponsors to support the theme of the exhibition. This was proof that the idea, not just of a bridge but an inhabitable one, was likely to appeal to a wide audience.

I was therefore delighted that the Royal Academy took on and imaginatively adapted this theme for their own exhibition. This bears testimony to their long-standing commitment to architecture and to be at the forefront of new and exciting ideas. Their proposal for a combined display of historic examples of inhabited bridges together with provocative ideas for a future London Bridge is a refreshing way to encourage a closer relationship between the past and the future which often seems so difficult to achieve.

The ideas competition attracted a wide variety of proposals. It was a privilege to be a member of the jury panel, to hear at first hand what each architect sought to achieve and to engage in dialogue with them and others on the panel.

In 1995, I set up the Thames Advisory Group to look at ideas to improve London's great river and the architecture of its banks. Much progress has been made and a wider acceptance of the need for change is slowly turning ideas into action on many fronts. I see this exhibition as part of that process.

Although many will seek to maintain the status quo *and do not wish to see familiar vistas altered, there is a demonstrable need and overwhelming support for improvement. There are cases where great new projects of lasting value and grand scale should be built that significantly improve the quality of life in our cities. An inhabited bridge must be one such case: an opportunity to create context as well as to respond to it and offer new, inspiring and breathtaking views of a great river whilst improving the quality of the city's urban fabric.*

This is the question which the Royal Academy puts before us. In doing so it becomes clear, yet again, that so much which is of special relevance to Britain is at once part of a great European-wide tradition.

Will such a bridge be built? That is the question for all London to answer.

The Rt. Hon. JOHN GUMMER MP

Foreword by the Director of the Musée national d'art moderne – Centre de création industrielle, Centre Georges Pompidou

From its inception in 1977, the Centre Georges Pompidou in Paris has deliberately presented its visitors (seven million annually) with a multidisciplinary approach to twentieth-century art and culture. Hence, architecture has formed the subject of over 160 exhibitions. During this period Jean Dethier has contributed greatly to this on-going, innovative policy which has attracted large audiences. Since the early 1990s, he has given particular preference to exhibitions focused on the city and urbanism, on urban design and cultures.

As architecture has gradually established a new 'right to be heard' in museums over the past twenty years, it is important to recognise that the city, as a major component of modern culture, has been almost entirely neglected as a legit imate theme for exhibitions.

It was in order to counteract this strange, paradoxical cultural amnesia that Jean Dethier pioneered the first major interdisciplinary exhibition in Europe on the theme of 'the modern city' which was shown at the Centre Georges Pompidou in 1994 and then in Barcelona and Tokyo in 1996. It was with the same innovative intent that in 1991 he began work in collaboration with Ruth Eaton on an exhibition dedicated to a building type which frequently represented the spirit of urban vitality: the inhabited bridge in the past, present and future.

The exhibition Living Bridges: the inhabited bridge, past, present and future, *which was thus initiated in Paris by the Centre Georges Pompidou and completed by the Royal Academy of Arts in London, constitutes an exceptional event on several counts. It explores a fascinating yet little known topic which has never been studied before in any depth and hence can be considered a trailblazing event. The exhibition presents a creative interpretation and a vigorous application of that interdisciplinary approach so central to the Centre Georges Pompidou's mission since, by definition, the inhabited bridge assumes a synergy between three disciplines which are rarely reconciled: engineering, urban planning and architecture. The exhibition provides the possibility of a coherent, on-going study of the many different stages of the evolution of the building type, from its medieval origins to its most contemporary manifestations. In addition, it has very real implications for the issues of today. By proposing to resurrect the inherited wisdom of inhabited bridges it is able to establish constructive proposals for the modernization of the building type relevant to the requirements of cities and citizens in the future.*

In this respect, the link between the exhibition and the international architecture competition organised by the Royal Academy in order to encourage the construction of a great new inhabited bridge over the River Thames, in the heart of London itself, represents a stimulating cultural event and allows seven very different entries to be presented to the public, to politicians and to developers.

An organic relationship between an architectural exhibition and an architecture competition in order to realize an exemplary project has twice been explored at the Centre Georges Pompidou by Jean Dethier. In association with his exhibition, Down to Earth, *he encouraged a development of 72 experimental, ecologically-sound houses to be built in 1983 by the State in the new town of Île d'Abeau near Lyon. In 1989, in the wake of the exhibition* Châteaux Bordeaux, *which invested the architecture associated with vineyards with a new value, a private company constructed an innovative winery for Château Pichon-Longueville at Pauillac in the Medoc. These two successful enterprises prove that it is possible for museums and cultural institutions to become directly involved in the debate on the future of architecture and urban design. This form of civic action, still very rare, reflects the mission of the Centre Georges Pompidou as enunciated in its law of foundation passed by the French Parliament. I am delighted that, in the context of the exhibition* Living Bridges, *the Royal Academy can share this dynamic cultural approach with us, in close cooperation with the British Government represented by the Secretary of State for the Environment.*

Such projects permit cultural institutions to act less passively in respect of the destiny of our cities, to participate in the debate concerning the future of our common urban environment.

I am delighted that these visionary ambitions could be realized within the context of the first collaboration between two major European cultural institutions: the Royal Academy of Arts and the Centre Georges Pompidou. I wish to thank all those who have worked so hard to realize this ground-breaking enterprise.

GERMAIN VIATTE

Paris, August 1996

Exhibition Sponsors' Prefaces

As trustee of one of the most famous inhabited bridges, medieval London Bridge, the Corporation of London is delighted to be involved in this inspirational exhibition charting the history and future of living bridges.

As local authority for the City of London our prime role is to support the City as the world's leading international financial centre, from looking after its townscape to devising positive policies to ensure its economic well-being. However, our other responsibilities extend far beyond the Square Mile. We are, for example, the third largest sponsor of the arts in this country, we provide internationally sought-after museum, library and art gallery facilities and protect over 10,000 acres of precious open spaces.

This sponsorship is a rich addition to our arts portfolio and totally in keeping with our deep involvement with London as a world capital. As we continue to maintain the ancient river crossing at London Bridge, and also run the three other Thames bridges that cross into the City, we look forward with anticipation to the River Thames flowing into the twenty-first century.

MICHAEL CASSIDY
Chairman
Policy and Resources Committee
Corporation of London

Générale des Eaux is delighted to co-sponsor the Living Bridges *exhibition. For more than a century we have been France's leading water company and we have developed unrivalled expertise in the engineering and operation of water services world-wide. Like living bridges through the ages, we have developed in size, strength, diversity and expertise. Our 2,500 operating companies employ 215,000 people. They range from water, energy and waste management to construction, civil engineering, transport, health care and telecommunications. In the United Kingdom alone, some fifty group companies employ 24,000 people.*

Yet, despite our size, Générale des Eaux remains essentially a group of small- and medium-sized companies which have grown from, and retain very close ties with, the areas in which they operate. So, in common with the great living bridges of the past and present, Générale des Eaux has both successfully withstood the test of time and adapted to change.

JEAN CLAUDE BANON
Chief Executive
Corporate Affairs
Générale des Eaux UK

Competition Sponsor's Preface

Thames Water is proud to be sponsoring the Habitable Bridge Competition culminating in the Living Bridges *exhibition.*

It is the first time we have worked with the Royal Academy of Arts and, as the River Thames is central to our business, we are particularly delighted to be involved in this exciting project for London.

We care a great deal about London and the River Thames. Thames Water is investing heavily in improvements above and below ground in London. Above ground, we are currently spending £1 million a week to ensure that the quality of the water returned to the River Thames is among the highest in Europe. It is now the cleanest metropolitan river in the world and has witnessed the return of otters to its waters for the first time in thirty years.

We hope that you enjoy visiting this fascinating exhibition as much as we have enjoyed investing in the Habitable Bridge Competition and in the future of London and the River Thames.

Sir Robert Clarke
Chairman
Thames Water Plc

Fig. 1, Nigel Coates, *Sketch design for the exhibition*, 1996, mixed media . © Nigel Coates

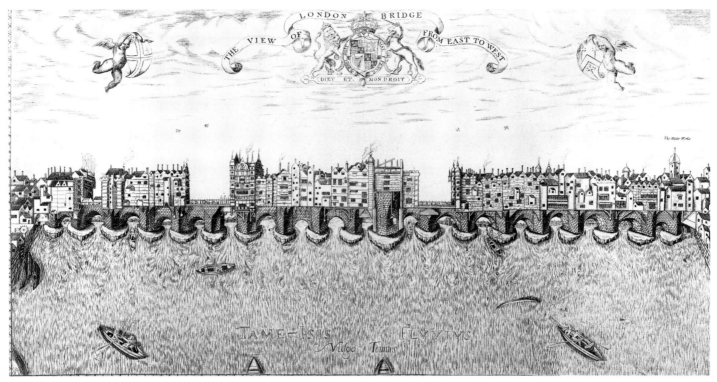

Fig. 2, John Norden, *The view of London Bridge from east to west*, c. 1594, engraving. © Museum of London

The Bridge and the City

Peter Murray

A 'river of time' flows through the Royal Academy (fig. 1). At its source stands a magisterial, 1:200 scale model of Old London Bridge, its thirteenth-century arches surmounted by a gallimaufry of architectural styles and structures. Set against a skyline of London in about 1600, the model reflects John Norden's contemporary description: 'It is adorned with sumptuous buildings and stately and beautiful houses on either side; inhabited by wealthy citizens and furnished with all manner of trades, comparable in itself to a little city' (fig. 2).

The 'river's waters flow on beneath Florence's Ponte Vecchio, completed in 1345 and still one of that city's most famous landmarks.

Clearly visible are alterations carried out by Grand Duke Cosimo I de' Medici in 1565, when he added a corridor designed by Giorgio Vasari to provide a private route from the Uffizi to the Palazzo Pitti. Next come two of Paris' bridges, one built, the other an extravagant but unrealized project by du Cerceau, connecting the Île de la Cité with the two banks of the River Seine to create a homogeneous urban plan in which the river was ignored and the pedestrian unaware of the river running beneath his feet. The Parisians demolished all their multifunctional, multi-layered inhabited bridges in the eighteenth century in order to make way for bridges dedicated solely to vehicular traffic and the creation of vistas of the river and the cityscape beyond.

Fig. 3, *Model of the Rialto Bridge, Venice, designed by Antonio dal Ponte in 1588*, scale 1:200. Modelmakers: Andrew Ingham and Associates. Royal Academy of Arts

The jewel-like form of the single span Ponte di Rialto in Venice – an engineering triumph of its time – was designed by the appropriately named Antonio dal Ponte (fig. 3), chosen in preference to Andrea Palladio. Subsequently commemorated in Canaletto's many pictorial reconstructions, Palladio's design is now brought to life in the exhibition (fig. 4), as are William Bridges' imaginary designs for a bridge across the Avon Gorge, outside Bristol, with warehouses and hoists driven by wind power and Sir John Soane's monumental triumphal bridge, his entry for the Royal Academy's Gold Medal in 1776. The last inhabited bridge to be built was Pulteney Bridge in Bath in 1770 (fig. 5). Since then the concept of an inhabited, multi-functional bridge has intrigued architect and engineer alike. However, from Lutyens' Dublin Art Gallery and Holden's replacement for Tower Bridge to Melnikov's bridge-garage for Paris and Jellicoe and Coleridge's design for a multi-storey bridge across the River Thames at Vauxhall, the building type has remained on the drawing board.

The relevance to the modern city of these habitable bridge projects has largely gone unrecognized. Many of the factors which originally encouraged the development of inhabited bridges have disappeared. No longer do bridges have a defensive role requiring gates and accommodation; water mills and toll houses have gone as has pressure on land within the medieval walled city which forced citizens to use every available space within their protective embrace. The sanitary advantages of living directly above flowing water were superseded by the public sewerage system.

However, attitudes are beginning to change. Today there are serious projects for inhabited bridges in London, Dubai and Rome, and proposals, with longer odds, for many other cities around the world. While fascination with this building type for architects has never flagged, it has taken some two hundred years for it to regain a wider acceptance.

The pioneering work of Jean Dethier at the Centre Georges Pompidou has revealed an identifiable typology. By tracing the origins of inhabited bridges, the causes for their inception and the reasons for their ultimate demise, Dethier and his colleague Ruth Eaton have created the theoretical basis upon which it is possible to assess the relevance of such structures today. The exhibition *Living Bridges* carries Dethier and Eaton's work forward: the evocative environment designed by Nigel Coates – where the 'river of time' wends its way through medieval city and neo-classical landscape towards the Millennium – allows the visitor to address the role of the inhabited bridge in the modern city, and in particular in London. The final span across the 'river of time' – by now transformed into the River Thames – presents the climax of the exhibition: the designs for a new inhabited bridge for London.

These proposals are the result of an architectural competition for one of two sites identified in the Government Office for London's 'Thames Strategy' as an area requiring greater accessibility across the River Thames. The Thames Strategy considered planning issues affecting the River Thames from Hampton (to the west of London) to the Thames Barrier at Woolwich, East London. The study considered matters relating specifically to the river – water-borne traffic, water supply and pollution control – as well as the relationship between the two banks, the effects of such development and the impact of the immediate bank on its hinterland. For a new inhabited bridge to function successfully, it had not only to engage in the problems of spanning a given section of the river but also to consider the impact of its presence upon both the adjacent urban zones beyond its docking point on each bank and on the tendrils of improved communication which it would push out into the heartland of the city.

The first site identified by the Thames Strategy was the stretch of river between Bankside, the site of the new Tate Gallery of Modern Art on the South Bank, and St Paul's Steps on the North Bank. The second lay adjacent to the South Bank Centre and, on the northern side, Aldwych (see figs 220, 221). The two sites form part of a 'cultural crescent' which runs along the southern edge of the River Thames from Butler's Wharf in the east to County Hall in the west. It consists of a chain of cultural and tourist facilities that promises to create one of London's most vibrant quarters. The area enjoys the environmental benefits of river views but suffers from the river's devisive power. The River Thames is wide: at about 250 metres across it constitutes a considerable barrier, both physical and psychological, to movement between the two banks. The distance across the river is exacerbated by traffic, vehicular access and level differentials. The North Bank succeeds in generating the intensity of use commensurate with a central urban location; the South Bank does not. Indeed, areas such as Covent Garden are now so heavily visited that it is deemed necessary to find ways of attracting alternative activity, particularly tourism, southwards over the river.

An inhabited bridge, while able to fulfill the function of a purely vehicular bridge, has two additional assets: it can extend the intensity of activity generated on each bank, and become a destination in its own right. It also has the potential for being self-financing. This was the case with Old London Bridge which generated so much income from tolls and rent that it left an inheritance in the form of the Bridge House Fund. This produces an income of some £10 million per annum for the City of London. Although initially intended solely to underwrite the cost of repairs to bridges within the City of London, its endowment is so significant that it supports a range of charitable enterprises.

In order to assess the financial viability of a modern inhabited bridge in London, the Secretary of State for the Environment and Minister for London, the Rt. Hon. John Gummer MP asked the international management consultants KPMG to carry out a feasibility study as to the potential uses, costs and income of an inhabited bridge linking the North Bank of the River Thames at Temple Gardens just south of Aldwych, to the London Television Centre, east of the Royal National Theatre. The KPMG study found that the bridge would be commercially viable as a stand-alone development without the need for any injection of public funds. This assessment was supported by the initial reactions of a number of developers. Several factors affect the financial viability of a bridge of this type. First the location is of paramount importance since it has a direct impact on rents and yields, and hence upon the value realizable from the bridge. Secondly, the combination of uses proposed for the space, including restaurants, shops, offices, housing, hotels, and leisure and cultural activites, together with the current levels of demand in each of these areas, is highly pertinent. It was recognized that, with potential development space

ranging from 15,000 square metres to 35,000 square metres, the latter would clearly be the most profitable but might also prove the most difficult to handle in terms of the scale of the resultant structure. Finally, it was understood that it would be essential for public support to be gained if the bridge were ever to be built.

Seven leading European architects were invited to propose solutions to the problem of designing a structure which would provide sufficient commercial space while protecting some of the finest views in London. The architects represented a spectrum of contemporary architectural styles: Zaha Hadid and Daniel Libeskind represented deconstructivism; Future Systems and Ian Ritchie that of high-tech; Leon Krier and Antoine Grumbach provided a Continental European dimension, and Branson Coates presented a more individual narrative style.

There are no contemporary built prototypes for the design of an inhabited bridge; each architect had to reinvent the form. The functional requirements of today as well as the technology are very different from those with which a medieval or a classical designer was confronted. Whereas the medieval inhabited bridge ignored the natural beauty of the river which it spanned, this becomes a major asset for its contemporary counterpart. Likewise, the short spans of older bridges are no longer technically necessary nor applicable to a site where, in order to ensure the safety of river traffic, a massive span is required which provides a design challenge for the architect and the engineer.

The realization of an inhabited bridge to span the River Thames between Waterloo Bridge and Blackfriars Bridge will depend on a positive public response to the designs, as well as on a developer who is willing to risk reinventing a building type last attempted over two hundred years ago. While this will surely require a great leap of faith, the project has a precedent in the development of air rights over railway stations. Pioneered by the developer Stuart Lipton, offices have recently been constructed over the tracks at Victoria Station, Cannon Street Station and Liverpool Street Station. Indeed, one of the buildings at Broadgate, developed by Lipton and his erstwhile partner Godfrey Bradman, is in effect an inhabited bridge above the railway at Liverpool Street Station. Exchange House, supported by a giant catenary arch spanning a massive 78 metres across ten railway tracks, accommodates 370,000 square metres of office space.

The aim of this exhibition is to create a debate around a recently rediscovered building type, the inhabited bridge. This debate will investigate the issues, hear the arguments and, we may hope, see the construction of an inhabited bridge within the heart of London.

Fig. 4, *Model of a design for the Ponte di Rialto, Venice, proposed by Andrea Palladio but never built*, scale 1:200. Modelmakers: Andrew Ingham and Associates. Royal Academy of Arts

Fig. 5, *Model of Pulteney Bridge, Bath, designed by Robert Adam (1770-73)*, scale 1:200. Modelmakers: Andrew Ingham and Associates. Royal Academy of Arts

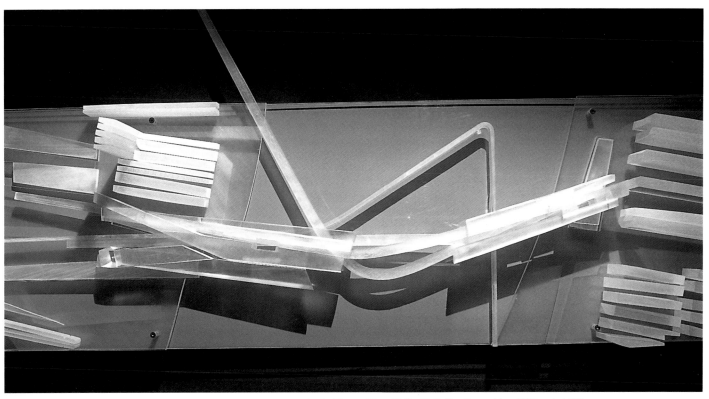

Fig. 6, Zaha Hadid, *An inhabited bridge for the twenty-first century: entry for the Thames Water Habitable Bridge Competition*, 1996, scale 1:200.

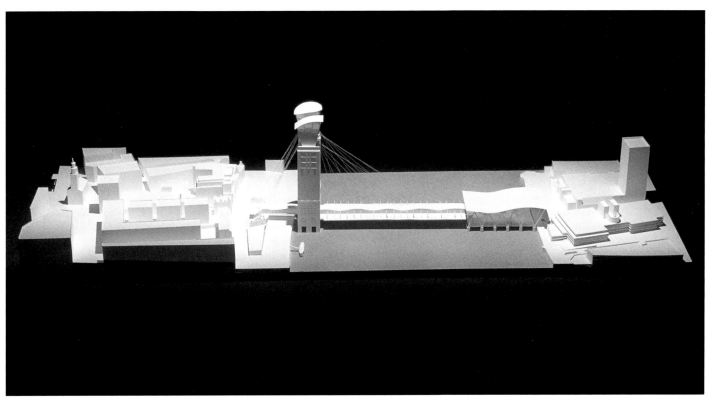

Fig. 7, Antoine Grumbach, *An inhabited bridge for the twenty-first century: entry for the Thames Water Habitable Bridge Competition*, 1996, scale 1:200.

Inhabited Bridges past, present and future

An interview with JEAN DETHIER

An 'inhabited bridge' as a building type has, until recently, escaped the attention of architects, planners and historians. Is this perhaps in part due to the absence of a precise definition of the term?

Neither French nor Italian provides a more precise definition than the English term used in the question: 'inhabited bridge', 'pont habité', 'ponte abito'. While implying a sense of animation the term does not make any reference to the distinct physical form of such bridges. It is only in German, where the building type is defined as 'die Überbautenbrücke' (bridge which is built upon), that a distinct category of bridge is clearly defined.

How would you propose to define the building type?

An inhabited bridge – in addition to its primary function of surmounting natural or man-made obstacles, be they rivers or canals, railways or motorways – serves as an organic link between two urban areas by connecting them to each other with a development of buildings erected on the bridge deck to form permanent accommodation for various social and economic activities (fig. 8). Thus every inhabited bridge consists of two elements: the platform that spans the obstacle and an architectural superstructure. In contrast to a purely vehicular bridge, the inhabited bridge provides a continuity within the urban fabric that is not only social and economic but also cultural, emotional and symbolic at a point where a natural break would otherwise exist. Indeed, it is both seductive and functional.

In order to understand more clearly the impact of this exceptional building type upon the urban fabric, most notably in Europe, three questions have to be addressed. How and why did Europe evolve such a logical, convenient and harmonious concept, and what form did its development take?

Why did the building type eventually disappear from current town-planning practice and why do historians, planners and architects continue to neglect it? And is it possible to provide a contemporary rationale for the building type in order to argue for its introduction into current and future town planning?

When reviewing the historical development of inhabited bridges, it is possible to identify four phases. Inhabited bridges appeared in Europe during the Middle Ages – in the eleventh or twelfth centuries, depending on the country or region concerned – and enjoyed their heyday between the late Middle Ages and the end of the sixteenth century. They went into decline during the seventeenth century and, apart from Pulteney Bridge, in Bath, built in 1770 (see pp. 72, 73, figs. 93-97), most were in desuetude and demolished during the course of the eighteenth century (fig. 9). A provisional review suggests that over a hundred inhabited bridges were built in Europe between the Middle Ages and the Age of Enlightenment; only ten or so survive today. Although no public or private investment in these bridges has taken place since the eighteenth century, the building type has continued to fascinate architects.

What appeal did the building type hold for architects during the nineteenth century?

Hundreds of projects have seemingly come off the drawing board over the past two hundred years, but only a handful have been realized: two near Paris (Noisiel and Reuil-Malmaison); one in London (Tower Bridge) and one in California by Frank Lloyd Wright. It would appear that lack of any institutional interest in the building type has meant that each architect has had independently to rediscover the original inhabited bridge.

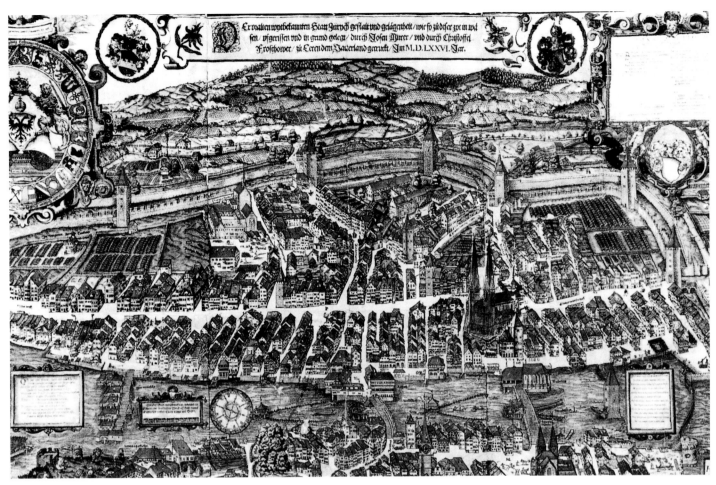

Fig. 8, *The City of Zurich* (detail), 1576, wood engraving. Baugeschichtliches Archiv der Stadt Zürich

Fig. 9, *Old Tyne Bridge, Newcastle, in a ruinous state,* 1772, engraving. Newcastle City Libraries and Arts

Fig. 10, Jules Saulnier, *Power plant over the River Marne*, *Meunier Chocolate Works,* Noisiel, France, 1869- c. 1888. Association Noisiel Ville d'Histoire. The external façade is clad in polychrome brick making a chequer-board pattern and was non-load-bearing. The mill was renovated in 1966 as the Nestlé company headquarters.

In England, these imaginative projects ranged from the innovative theatrical bridge proposed by William Bridges in 1793 for crossing the 60 metre-high Avon Gorge outside Bristol with the inhabited structures placed beneath the deck (see p. 82) to the construction of Tower Bridge, in London, at the end of the nineteenth century – a structure whose pseudo-habitable parts contained machinery and a lookout, rather than residential or commercial accommodation (see pp. 98, 99). More overtly engineered were the novel proposals by Mosley in 1843 for an art gallery over Waterloo Bridge (see pp. 86-88); by Frank Lang in 1861 for a concert hall and winter gardens inspired by the great stores of botanical gardens, also to be sited over Waterloo Bridge; and by Baldwin for a suspension bridge over the River Thames decked out with small market stalls to become a 'bazaar-bridge' (see p. 88). In France during the same period, three projects demonstrated the persistent fascination with inhabited bridges: Baltard proposed two parallel bridges linked by an artificial island to accommodate public buildings giving on to a central square across the River Saône at Lyon in 1823 (see pp. 84, 85); Eiffel failed to convince the authorities of the virtues of a 300-metre long glass-and-iron bridge with a vast 100-metre glazed hall above the Pont d'Iéna for the Exposition Universelle in 1878 (see p. 90); and Keck, in 1899,

envisaged a commercial and cultural building in the same materials, across the River Rhine at Basel (see p. 90). In addition, there was the remarkable revision of the medieval mill-bridge designed by Jules Saulnier and built in 1869 for Meunier's chocolate works at Noisiel, south of Paris (fig. 10). Designed to accommodate the turbines used to generate power for the factory, the five storey cast-iron structure rested on four piers across an arm of the River Marne. Beyond these two countries, Amsterdam received a proposal in 1848 from the engineer Galman to create a huge multifunctional bridge across the Ij outside the city as a strategic key to the development of the empty opposite bank (see p. 89), while the Ponte Vecchio, in Florence, might have been transformed into a cast-iron and glass 'galleria', had plans by Martelli and Corazzi been implemented (see p. 88).

How have architects responded to the challenge of the inhabited bridge in the twentieth century?

Over a hundred projects for some twenty European cities in twelve countries have been conceived during the twentieth century. The majority of these projects have been produced by architects of international repute and in a range of different styles, from classical to high-tech, from Expressionism to Constructivism, from Rationalism to Megastructuralism, from art deco to pop, from Arts and Crafts to Post-Modernism, and from deconstructivism to contextualism. At the beginning of the century, Collcutt and Lutyens provided classicizing solutions for bridges for London and Dublin respectively (see pp. 92, 93), while in 1920 Raymond Unwin proposed an Arts and Crafts inhabited bridge for Letchworth (admittedly over railway lines) and Berlage in 1907 a grand brick entrance in his own lyrical variant of Expressionism for a bridge leading to his new urban development in south Amsterdam. On the occasion of the Exposition Internationale des Arts Décoratifs held in Paris in 1925, the Russian Constructivist Melnikov proposed a garage-bridge eight storeys high across the River Seine (see p. 97) and during the 1930s Le Corbusier proposed for several cities, including Algiers and Rio de Janeiro, variants of his concept of the 'bridge-town'. His immense bridges, creating ribbons across the urban terrain, would support motorways below which would lie a continuous sequence of structures. Although his vision was only very partially realized by a disciple in Algiers in the 1950s, Le Corbusier's visionary solution to the inhabited bridge foreshadowed the architectural megastructures which emerged world-wide during the 1960s and profoundly affected the morphology of the inhabited bridge. The French architect Friedman proposed similar structures for Monaco (1959) and also as a link between France and England (1963); the Austrian Abraham likewise wanted to bridge the Channel with a 'mega-bridge' – rising to thirty storeys – in 1966, and Kenzo Tange, with the Metabolist Group, presented the apotheosis of this megalomaniac approach in an inhabited bridge marching across Tokyo Bay. The only challenger to Tange is Gaetano Pesce – Italian artist and architect – with his 600-metre long Pont de l'Europe across the River Rhine near Strasbourg, designed in a pop art style. At the other end of the spectrum, Italian architects contemporaneously produced the twentieth-century riposte to the rural inhabited bridge in the form of the Pavesi restaurants sprung gracefully across the autostradas. German architects were also intrigued by the possibilities presented by the building type, with proposals being made for Düsseldorf by Böhm and by Bunsman for a new Parliament building across the River Rhine at Bonn, while in 1968, the American architect, Louis Kahn conceived a convention centre thrown across a Venetian canal. Gregotti turned his attention to Southern Italy, with a blueprint for a new university whose accommodation would be provided by inhabited bridges laid out across the landscape.

More recently, the challenge of the inhabited bridge has been met in relation to bridging railways and motorways as well as rivers. Two French architects Gregotti and Grumbach have both proposed to link the two banks of the River Seine with a bridge, as part of a World's Fair urban design project for 1989. Among President Mitterand's 'grands projets' was a proposal by Jean Nouvel to house the Bibliothèque Nationale de France in five buildings arranged in a fan shape, the middle one being extended over the River Seine to create

a cultural and symbolic relationship between the Left Bank and the Right Bank. In 1988, Rob Krier undertook an urban renewal project for Amiens which used inhabited bridges as a unifying thread within the city centre (see pp. 108, 109). Elsewhere in France, Rem Koolhaas, as part of his extensive 'Eurolille' project for Lille, included a convention centre on a bridge across the railway tracks, a solution to the use of such 'free' urban land which has also been addressed by Mario Botta across twenty railway tracks outside Zurich Central Station. And an inhabited bridge across a motorway, rather than railway tracks, has been created by the urban planner Michel Richard as a way of mitigating the brutal rupture of Reuil-Malmaison and its new extension along the River Seine – Reuil 2000 – by an urban freeway; here the inhabited bridge, the Place de l'Europe, completed in 1993 transforms the traditional linear structure of an urban street into a one-hectar square piazza complete with cafés, shops and six storeys of offices. The Place de l'Europe, is the first inhabited bridge in the twentieth century to expand the traditional limitations of the building type by creating a square – not a street – and by providing office accommodation.

Which geographical area has been most favourable to the development of the inhabited bridge?

All the evidence suggests that the inhabited bridge is a concept peculiar to the towns and cities of Europe, though this does not apply to the entire continent. No significant traces of the building type exist in southern Europe (Spain, Portugal, Greece), Scandinavia or eastern Europe. The three European countries principally concerned – England, France and (northern) Italy – formed the transverse axis along which the vast majority of inhabited bridges were constructed over a period of eight centuries. Several countries to its north were also involved, albeit to a lesser extent – notably, in descending order of importance, Germany, Austria, Switzerland, Belgium and Holland.

Has the idea of the inhabited bridge spread beyond the borders of Europe?

The United States of America discovered the building type comparatively late. Three successive phases can be distinguished there. During the 1920s, two architects – Mullgardt in San Francisco and Hood in New York – put forward proposals for inhabited bridges of unprecedented size (see pp. 94, 95, figs 137-140). Both bridges were, in fact, conceived as skyscrapers, to be constructed across San Francisco Bay and the Hudson River respectively and designed to serve as gigantic piles supporting suspension bridges onto which complementary buildings would have been grafted. These megalomaniac structures were intended to house or provide places of work for between 25,000 and 100,000 people. However, the economic crisis of 1929 banished all hopes of erecting inhabited bridges on such a vast scale. The second phase, from the late 1930s to the immediate post World War II period, saw only Frank Lloyd Wright addressing the question of the inhabited bridge, both in the form of a private

house suspended over water (fig. 11) and in the form of a civic centre, placed not over water but across two expressways (The Civic Center, Marin County, outside San Francisco). The third phase, dating from 1970 onwards, displays wide divergences from the original model in addressing itself to rural sites – the 'cultural bridge' designed in 1979 by Michael Graves for Fargo and Moorhead (see pp. 104, 105, figs 157-159) – as well as urban ones (the Northern Avenue Bridge, Boston, Massachusetts, 1996, a proposal by Wellington Reiter; see pp. 124, 125, figs. 207-211).

Do inhabited bridges exist in any non-Western civilizations?

Examples in the Middle East and in Asia seem to be rare. There appear to be isolated exceptions unrelated to any continuous tradition, but more detailed research has yet to be undertaken in this field. To date, the most elaborate example known to us is the Iranian barrage bridge in Isfahan. Constructed in the seventeenth century, it served a number of functions: a passage over the Zayandeh Roud, a dam, and a place for guests to view acquatic festivals. However, despite the sophistication of its architecture and engineering, it lacks the provision of commercial, residential or cultural facilities which would have allied it to European inhabited bridges (fig. 12).

What were the main functions of European inhabited bridges?

It appears that the dominant function of most inhabited bridges was commercial. This state of affairs frequently evolved from relatively modest shops that gradually gave way to establishments of an increasingly prestigious nature.

This process is well illustrated in the case of the Ponte Vecchio, in Florence, where butchers were compelled to surrender their stalls to

Fig. 12, *The Hasan-Beg Barrage Bridge, Isfahan, Iran*, 17th century, engraving. Bibliothèque des Arts Décoratifs, Paris, Collection Maciet

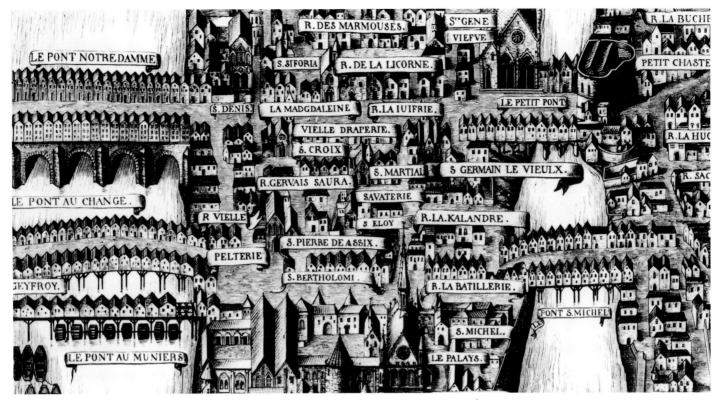

Fig. 13, *Copy after the Plan de la Tapisserie (detail) showing the inhabited bridges spanning the River Seine from the Île de la Cité, Paris*, c. 1540. Paris, Musée Carnavalet

jewellers (see pp. 62-65). Similarly, a royal decree of 1141 obliged all the money-changers in Paris to concentrate their activities on the Pont au Change (see pp. 55-57). This process tended to invest bridges with particular prestige and enhance their status within the urban structure. More generally, the bridge lined with shops can be regarded as the ancestor of our modern commercial centre or shopping mall. In 1515 the buildings on Pont Notre-Dame in Paris were rebuilt in order to provide a continuous, regular row of shops incorporating display windows, an important innovation designed to catch the customer's eye. The bridge also asserted its modernity and created a secure environment by installing oil lamps for illumination at night. The need to safeguard their stock also prompted tradesmen to establish their living quarters close to their businesses. Thus we find apartments on one or more levels being built over the shops. By the end of the Middle Ages the inhabited bridges of London and Paris sometimes supported as many as four or five storeys of housing, thus becoming ordinary thoroughfares little different in scale from the streets on either bank. People crossing the Seine would not have been able to see the river, so tall and tightly packed were the buildings on its bridges (fig. 13). Only the bridges of Venice and Florence contrived to interrupt the rows of shops halfway across. While the

Ponte di Rialto in Venice incorporated a central archway from which to enjoy views of the city and the Grand Canal (see pp. 66-71), the opening in Florence's Ponte Vecchio was created for structural reasons to provide three arches for carrying Vasari's covered private passageway from the Uffizi to the Palazzo Pitti of 1565 (see pp. 62-65).

In the Middle Ages, bridges became 'inhabited' by virtue of the fact that they provided anchorage for mills making use of the river current beneath. In some cases, such 'mill-bridges' remained mono-functional, as for example at Zurich (fig. 8) and Tournai. Others gradually accumulated a multiplicity of functions, as was the case with several Parisian inhabited bridges (see pp. 52-61), Old London Bridge (see pp. 46-51) and, most notably, the Mühlendammbrücke in Berlin (figs. 14-16). The evolution of this bridge commenced in the thirteenth century, when four mills were attached to a wooden bridge structure. By the sixteenth century, the bridge had acquired shops, whose irregular architecture was rationalized into a harmonious row of buildings by Johann Arnold Nering at the end of the succeeding century (fig. 15); Nering also applied identical façades to the then six watermills on the downstream side. The bridge continued to function well into the nineteenth century (fig. 16).

Fig. 14, *Model of the centre of the city of Berlin showing the Mühlendammbrücke*, 1688, wood. Berlin, Stadtmuseum Berlin

Fig. 15, Andreas Ludwig Krüger, *Fischerstrasse, Mühlendamm, Berlin*, 1796, pen and ink. Berlin, Stadtmuseum Berlin

Fig. 16, *Spree Fischerbrücke, Mühlendamm, Berlin,* 1876, photograph by Albert Schwarz. Berlin, Stadtmuseum Berlin

Other bridges were constructed for or converted to strategic use. They could be integrated into the defensive system of the city walls, as was the case at Blois and, most strikingly, at Cahors, in south-west France (fig. 17), whose fourteenth-century bridge is punctuated by three tall towers that lend it a robust and spectacular architectural form. Alternatively, they contributed a more unusual defensive strategy by becoming barrage bridges that could, in the event of an attack, flood the land surrounding the city. Strasbourg bears witness to this form of strategic ingenuity, conceived in the seventeenth century by Vauban, Louis XIV's military engineer.

Apart from the commercial, industrial and military functions of inhabited bridges, they also provided space for religious structures. Why?

Towns were often founded on the banks of rivers because their waters fulfilled important functions in respect of industry and

Fig. 17,
Cahors,
View of the Pont Valentré,
19th century, lithograph.
Bibliothèque des Arts Décoratifs,
Paris, Collection Maciet

remarkable in that some were wholly financed by alms and even by papal and episcopal indulgences. The chapel-bridge, which seems to have originated in Italy, spread rapidly to France, England and Germany (see pp. 40, 41): the most celebrated surviving example is Avignon's Pont Saint-Bénézet (see figs. 37, 38). Other inhabited bridges extended this specific religious function to include monastic and medical institutions. The Ponte alle Grazie in Florence, still standing in the 1920s (fig. 18; see pp. 62-65), consisted of a series of individual convent buildings placed above each pier with a chapel at one end. Hospitals run by religious orders existed in both Nuremberg (the Heiliggeistspitalbrücke; see fig. 19) and in Paris (the Pont de l'Hôtel-Dieu, whose wards were demolished in 1769; fig. 20).

How do you explain the appearance and development of inhabited bridges during the Middle Ages?

As mentioned earlier, a large number of European towns and cities were founded on the banks of rivers because water constituted an essential element in the life and activities of a community. In most cases they developed on both sides of the waterway, which thus became a kind of barrier in the very heart of the populated area. Given that medieval towns were enclosed by walls for defensive purposes, building land within them was necessarily limited. Thus, as towns

navigation, commerce and defence. But rivers were also feared because of their disastrous floods, which were perceived as manifestations of the devil. In order to combat the forces of evil, the ecclesiastical authorities often built chapels and churches on bridges, since they were likely to be the first victims of nature's violent outbursts. In the Middle Ages, bridges gave rise to displays of piety that were

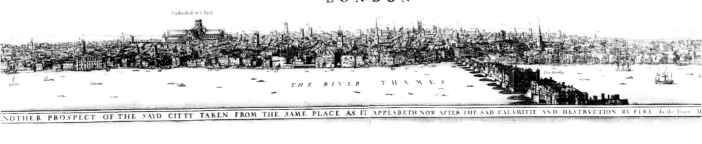

Fig. 21, Wenceslaus Hollar, *Double prospect of the City of London, before and after the Great Fire*, 1666, etching. © Guildhall Library, Corporation of London

grew and the population became denser, the pressure on the land often became intense. When the pressure became too great, the town would expand on the periphery and additional walls would be built to enclose the new buildings. Being situated ever further from the centre, however, the new districts would be commercially disadvantaged. As the political, social and economic conditions of the towns developed, the need to exploit the land at their very centre became apparent and the obvious course was to make fresh use of the bridges that linked the two banks. Being indispensable crossing-points between the two halves of the town, the bridges constituted prime commercial sites. Thus the increase in population density led to the urbanization of their bridges.

Which European inhabited bridges best illustrate the heyday of the building type?

I shall confine myself to citing one example from each of the three countries that have helped, historically, to formalize the concept of the inhabited bridge.

For centuries, Old London Bridge was the only bridge that connected the City of London with south-east England (see pp. 46-51). It was remarkable for four main reasons (fig. 21). In the first place, an extraordinary length of time elapsed between its construction in the twelfth century and its demolition in 1823. Secondly, it displayed an ability to regenerate itself again and again, despite the various disasters – floods, fires and frost – that beset it over the years. Thirdly, its size made it the longest inhabited bridge ever built in Europe. Finally, and most importantly, Old London Bridge accommodated a wide range of commercial and domestic functions, becoming in some senses a microcosm of the city itself.

The most remarkable examples in France were those of Paris (fig. 26), where, for several centuries, all four bridges connecting the Île de la Cité to the right and left banks were surmounted by some impressive and innovative buildings (see pp. 52-61). Often having structures

rising four or five storeys, and with the highest density of urban population, these inhabited bridges were quickly rationalized by the Parisians to form architectural ensembles and planned urban entities. This process was complemented by the very elaborate scenographic treatments of the bridges' internal spaces, which also confirmed them as the city's favourite venues for large-scale public festivals, processions and royal cortèges. Paris endowed its inhabited bridges with exceptional dignity, prestige and vitality.

Italy's finest example of an inhabited bridge remains the Ponte di Rialto in Venice, built in 1588 by the architect Antonio dal Ponte (see pp. 66-71). This bridge lends expression to three major innovations. In the first place, it optimizes its commercial function by incorporating three parallel footways flanked by four façades of shops, rather than the usual single axial roadway. It was also conceived as an object of extreme architectural refinement. While Paris's inhabited bridges had sought sophistication of design for their internal façades to the detriment of their river frontages, the Ponte di Rialto was designed and treated as an architectural whole, which would play an active part in bringing out the city's scenic beauties from every visual angle. By leaving a small, raised, open space half-way across, the bridge carries sophistication to the lengths of presenting the passer-by with a privileged view of the spectacle afforded by the Grand Canal. It is a masterly and well-considered compromise between the requirements of commerce, architecture, urban design, scenic beauty and social interaction.

What factors mainly contributed to the disappearance of inhabited bridges?

The history of inhabited bridges in Europe came to an end – only temporarily, I hope – in the eighteenth century. The reasons for their disappearance were economic, aesthetic and philosophical. Significant changes in military strategy, coupled with rapid economic growth, which with its expansion in commerce and trade, and the growth of urban centres, not only exerted pressure upon cities to

Fig. 22,
Canaletto,
*The Campo di Rialto and
S. Giacomo di Rialto, Venice,*
1740-1760, oil on canvas.
National Gallery of Canada/
Musée des Beaux-Arts
du Canada, Ottawa

spread beyond their confining walls but also increased the volume of traffic passing through their thoroughfares. Inhabited bridges created major constrictions to the free flow of such traffic and increasingly there were demands for their demolition. In addition, with changing attitudes to nature – from one which perceived it as a hostile force to one which recognized it as something to be contemplated for intellectual and emotional nourishment – the impediments of inhabited bridges to extensive vistas across urban landscapes became increasingly resented. The call for an uncluttered view was initially heard in Paris in the form of the virtually unencumbered Pont Neuf, begun in 1578, although it was especially towards the end of the eighteenth century that the full weight of the argument contributed to the inhabited bridge's destruction. In addition, it was during the eighteenth century that the training of engineers and architects became institutionally divorced. This professional segregation was to prove prejudicial to inhabited bridges precisely because they were products of a synergy between those two complementary branches of knowledge. With a few exceptions, notably Gustave Eiffel in the nineteenth century (see p. 90), 'bridge engineers' have never been attracted by the idea of encumbering their work with structures deemed by them to be parasitical. As for the architects of the eighteenth century, nearly all of them designed 'triumphal bridges', which – despite their sumptuous ornamentation – constituted ersatz inhabited bridges. These bridges no longer embodied any commercial or residential component; the turmoil of economic and social life had been replaced by a plethora of porticoes and colonnades (see pp. 76-81). It was during the eighteenth century, too, that rationalism in architecture favoured the rejection of programmatic complexity and

preferred to segregate functions in order to address them in isolation. Since the construction of inhabited bridges had most commonly been effected by a pragmatic approach, one can understand why the development of that process was checked in the eighteenth century (see fig. 50).

What has been the attitude of artists to inhabited bridges?

In Italy, France and England, artists testified to their admiration of inhabited bridges by immortalizing many of them, such that we can now appreciate the size and splendour of the bridges and grasp the extent of their vitality and urban importance. During the Middle Ages they appeared in illuminated manuscripts and, before long, on coins as well. In the seventeenth century the subject was taken up by painters such as Claude de Jongh in London. During the eighteenth century inhabited bridges became a subject frequently tackled by painters who, in the panoramic *veduta*, raised urban landscape to the level of a new artistic genre. Canaletto (see figs. 22, 89, 91), Guardi (see fig. 92), Raguenet (see figs. 61, 62, 74), Hubert Robert (see fig. 76) and Joli (see fig. 55) recorded the inhabited bridges of Venice, Florence, London and Paris. Their approach sometimes transcended the straightforward depiction of existing reality. Fascinated by some of Palladio's projects for Venice that never progressed beyond the architectural design stage, Canaletto produced a number of fanciful pictorial compositions that lent Palladio's visions a disturbing semblance of reality (see fig. 89). Many other eighteenth-century artists, including Piranesi, created imaginary inhabited bridges and thereby expressed a specific form of urban ideal. This approach was

Fig. 23,
J. M. W. Turner,
Blenheim Palace,
watercolour.
Birmingham Museums
and Art Gallery

Fig. 24,
Château de Fère-en-Tardenois showing the inhabited bridge in a ruinous state,
early 19th century, lithograph.
Bibliothèque des Arts
Décoratifs, Paris, Collection Maciet

to be perpetuated in the nineteenth century by Turner (figs 23, 119), Bonington (fig. 25) and Victor Hugo, or, within a rural context and treated as a picturesque ruin in early nineteenth-century lithographs (fig. 24). In the twentieth century both Paul Klee and Oskar Kokoschka have addressed the subject. The urban and social importance of inhabited bridges is thus attested by an almost unbroken tradition that spans the history of art. In the eighteenth century, when it was decided to demolish the last inhabited bridges in Paris, Hubert Robert realized that this was a salient event in the city's history: he devoted a series of paintings to these scenes of destruction, thereby imbuing them with poignancy and grandeur (see fig. 76).

Have any inhabited bridges been a determining factor in the execution of an overall town-planning scheme?

Inhabited bridges tended to create an organic relationship between two existing urban districts. But there are also instances where an inhabited bridge has been deliberately constructed to foster the growth of a new district whose development would otherwise have been hampered by its physical isolation from the rest of the city. In Paris,

Fig. 25, Richard Parkes Bonington, *The Grand Canal, Venice*, 1826, pencil and gouache. Trustees of the Tate Gallery, London

31

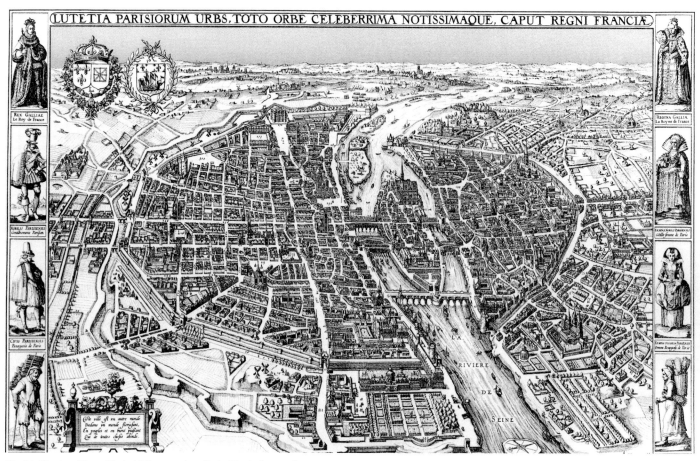

Fig. 26, Visscher, *Plan of Paris*, 1618, engraving. Paris, Musée Carnavalet

the Île Saint-Louis, upstream from the Île de la Cité, remained undeveloped until the beginning of the seventeenth century (fig. 26). In 1614 the authorities granted Christophe Marie permission to build, on condition that he first constructed two inhabited bridges linking the island to the left and right banks of the Seine, in order to create an organic relationship between this new quarter and the rest of the city (fig. 27; see also pp. 52-61). In Bath, the English watering-place, the 600-acre Bathwick estate, immediately adjacent to the city centre but separated from it by the River Avon, was opened up for development when, in 1770, William Pulteney commissioned the architect Robert Adam to build an inhabited bridge in the Palladian style lined with shops (see pp. 72, 73).

Do inhabited bridges exist outside an urban context?

It was the very density of towns and cities that justified adopting the idea of the inhabited bridge. There are, however, a few rare instances where celebrated buildings have been deliberately designed as bridges outside towns. Let us take two examples. The magnificent Château de Chenonceaux in the Loire Valley, takes the form of an inhabited bridge. This aristocratic residence, which dates from the sixteenth century, was in fact extended out across the river. The transposition of an urban model into a rural location was a private initiative. The bridge was not intended for public use but appears as an extravagant 'folly', a spectacular architectural cappriccio. On a more contemporary note there is Frank Lloyd Wright's exceptional and innovative Marin County Civic Center, which, although built on a spectacular site across expressways rather than a body of water, symbolically links the two suburban communities on either side.

Do you regard the Château of Chenonceaux and, more particularly, Frank Lloyd Wright's Civic Center as models for a new generation of inhabited bridges?

Yes, indeed. Our modern towns and cities are developing in an extremely heterogeneous manner and have little in common with the densely built-up cities of old. This applies particularly to the fragmented peripheries of our cities, where urbanization is dispersed and far less dense. Such developments render it essential to create new designs for suburban centres that will provide their communities with congenial and symbolic meeting-places. Since all these suburban areas are being drastically parcelled up and isolated from each other by motorways and other major roads, it is conceivable that, in this specific context, the inhabited bridge will discover a new role by spanning these modern obstacles and establishing a new generation of commercial, civic and cultural centres. Frank Lloyd Wright provided an architectural archetype that fulfils this requirement. As for the Château de Chenonceaux, it seems to me to represent the distant precursor of another set of issues. It was, in a sense, conceived as a 'holiday home' in the country. In contemporary society, where leisure pursuits have been democratized and leisure time is spent away from urban centres, it might be interesting to construct holiday villages on the countless disused bridges and railway viaducts that often traverse valleys in magnificent, unspoilt countryside throughout Europe. The idea of grafting holiday accommodation or hotels on to these thousands of abandoned but superbly situated bridges and viaducts strikes me as one possible means of reinventing the idea of the inhabited bridge in an extra-urban context.

Do you believe, therefore, that the future of the inhabited bridge lies outside the urban centre?

Certainly not! I think those two extra-urban contexts present possibilities from which new uses for the inhabited bridge might emerge. But that is a far cry from excluding the hypothesis that this type of structure might one day be reborn in the centre of cities or on their outskirts.

So what needs could be met by new inhabited bridges built in urban centres at the beginning of the twenty-first century?

Town planning of the kind largely practised in twentieth-century Europe has created major disfunctions in our cities, engendered deep dissatisfaction in their inhabitants, ravaged innumerable urban sites and squandered resources on a vast scale. Ever since the 1920s and 1930s, functionalism has dominated all other modes of thought. Enforced and excessive rationalization was the guiding principle. The Athens Charter, drawn up in 1933, reduces the complexity inherent in any city to an oversimplified concept that assigns precedence to four functions ('live, work, circulate, recreate') and leads the planner to believe that he is instilling order into, and mastering the problems of, urbanization. In the name of this essentially schematic dogmatism, the principles governing the segregation of urban functions and inhabitants have been applied wholesale. The creation of isolated institutional and functional ghettoes has established and generalized the urban and social fragmentation that threatens the cohesion of our cities. One of the new priorities of all planners should be to re-establish an organic link between the various arbitrarily separated urban entities: to revalidate the vital notion of urban complexity; to attempt to reconcile diverse and complementary activities in one and the same place; and to create spaces favourable to social interaction and places symbolically expressive of a desire to unite a city's inhabitants and their various occupations. We have lost the ability to conceive of, generate and manage the urban complexity that is so essential, and our cities are likely to die in consequence. This is precisely why the history lesson taught us by the history of inhabited bridges is important; for they created that very urban complexity so lacking in contemporary cities by super imposing several functions and concentrating them in the same spot.

Our modern bridges serve only one utilitarian purpose: they carry traffic. This restrictive practice constitutes a waste of resources which is doubly unacceptable because land is now at a premium in our urban centres. Every bridge comprises, by its very nature, a man-made site that should be used to both financial and social advantage as a base on which to graft elements of urban life, primarily by surmounting obstacles that prevent the city or its outskirts from forming

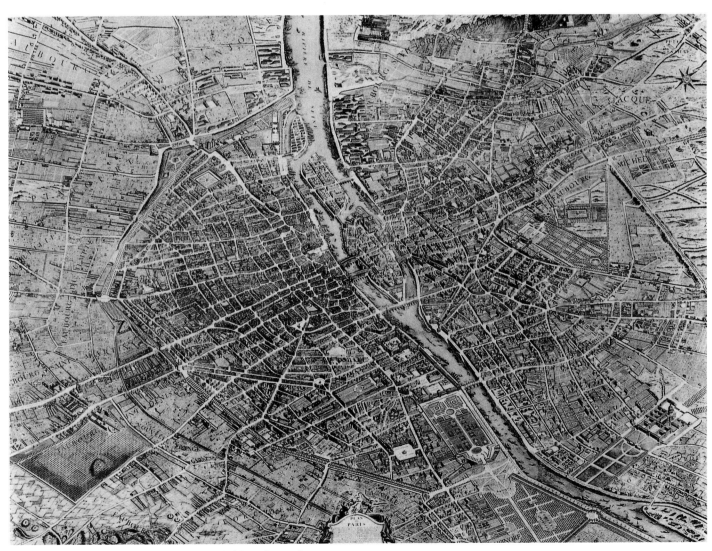

Fig. 27, *Paris, the Plan Turgot*, 1734, engraving. Paris, Musée Carnavalet

33

Fig. 28, SOM Architects and Engineers, *Exchange House, Broadgate, London*, 1988

a continuum; rivers, to be sure, but also, in the modern world, railways and urban motorways. This means that every city possesses many possible ways of reconquering space and revitalizing land that is presently being stifled by bureaucratic, unimaginative planning.

Would not the creation of these man-made sites be unrealistically expensive?

The economic study produced in London in 1996 on the initiative of the Secretary of State for the Environment discloses that the hypothesis for the self-financing of new inhabited bridges is thoroughly realistic. In the centre of London – and on the banks of the River Thames in particular – the cost of land and of buildings per square metre is so high that the overall cost of constructing and commercializing a bridge with buildings on it becomes reasonable by comparison. Indeed, such ideas have already been pursued specifically in London, not only in schemes for Blackfriars Bridge (see pp. 126-129), but also for the creation of new bridges at Bankside (see pp. 118-121) and at Embankment Gardens (see Cadman and Lipton, pp. 133, 134). Since the mid-1980s, London has been engaged in a debate focusing on the problems presented by the seemingly intractable problem of bridging satisfactorily a body of water the width of the River Thames. Proposals have been advanced by Richard Rogers (see pp.111-113), Terry Farrell (see pp. 126, 127), Richard Horden (see pp.120, 121) and Alsop and Störmer (see pp. 128-130). The issues raised by such schemes provide the focus of both the exhibition *Living Bridges: The inhabited bridge, past, present and future* and the Thames Water Habitable Bridge Competition which accompanies it (see Peter Murray, p. 135, and pp. 136-153).

Beyond the concern to readdress the traditional question of creating a new structure across a river – as is now happening in Boston, Cologne and Paris – in recent years politicians and decision-makers in cities had also explored the viability of constructing bridges over railway lines, as at Broadgate, London (fig. 28) where proven viability actually delivered an 'inhabited bridge', or spanning a town, as in Tschumi's proposal for Lausanne (fig. 29), or utilizing old railway bridges, as with Stephen Holl's scheme for New York City (fig. 30).

Are you confident that the idea of building new inhabited bridges will become a reality?

We cannot resign ourselves to seeing cities continue to develop on the basis of theories of planning initiated half a century ago that have too often seriously impaired our urban environment and way of life. We must therefore study more sensible and creative alternatives, so as to promote more dynamic and multifunctional projects, in which priority will be accorded to a plurality of urban functions and a new urban symbiosis. Of the various solutions to this problem, the inhabited bridge constitutes one of the hypotheses that merit serious and unprejudiced examination. It is a question, not of copying a historical model, but of drawing inspiration from its ethos and dynamism. During the 1930s Le Corbusier claimed that developers of the modern city must begin by renouncing the historical principle of the street. This dogma was naïvely applied by innumerable planners during the 1950s, 1960s and 1970s. Chastened by many tragic blunders, they finally realized that the street remains an essential feature

Fig. 29, Bernard Tschumi Architects, *Lausanne Bridge City*, 1988, model

of urban life and that it must be adapted to meet our new aspirations. The same applies to inhabited bridges. The time has now come to exhume them from oblivion, to comprehend their logic, to appreciate their merits and to devise for them new applications capable of remedying the defects and disfunctions of the modern city. Planners and decision makers must become less doctrinaire and bureaucratic, less functionalist and isolationist. They must henceforth assimilate a logic that is more pragmatic and all-embracing, more multivalent and humane. If they do, the inhabited bridge will become a model destined for a new future.

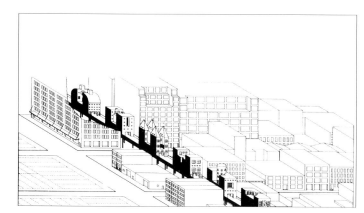

Fig. 30, Steven Holl, *Bridge of Houses, New York*, 1979-82, pen and ink. The inhabited bridge lies above an abandoned elevated rail link in the Chelsea area of New York City between West 19th and West 29th streets.

From
Medieval Times
to the
Eighteenth Century

Fig. 31, *Newcastle upon Tyne*, 1754, engraved for the *Universal Magazine*. Newcastle City Libraries and Arts

Medieval Bridges

Medieval Europe enjoyed an abundance of inhabited bridges; France alone possessed some thirty-five examples. This phenomenon was encouraged by the presence of defensive walls, which concentrated urban development into relatively constrained areas. Any vacant land was at a premium, including the open space which lay above a bridge linking a town's river banks, and thus presented development possibilities. While the availability of water power encouraged the construction of mills (see fig. 35), other types of building fulfilling a variety of functions – residential, religious, civic and commercial – also took their place on the area of passage across the water.

The trades plied upon these bridges varied widely, from grocers, butchers and blacksmiths to higher value trades such as goldsmiths, jewellers and money changers. Initially, tradesmen appear to have set up their shops without the sanction of the authorities as temporary kiosks or stalls. Such was the case, for example, in Venice where the magistrates responsible for the bridge decided to intervene in order to regularize the commercial activity from which they would obtain financial benefits (see pp. 66-71). Although more prevalent after about 1500, a few of the inhabited bridges encouraged the concentration of 'luxury' trades, often at the instigation of the ruling authorities. This development occurred in the case of the Grand Pont (later Pont au Change) in Paris, where, as early as 1141, the king had enjoined the money changers to concentrate their activities, regulated by the Crown, exclusively on that bridge. The money changers soon attracted related commercial activity such as goldsmiths and jewellers (see pp. 55-57, figs 62-65).

While specific case studies of the inhabited bridges in London, Paris and Florence are given elsewhere in the catalogue (see pp. 46-65), a brief review of three other examples is given below: Newcastle Bridge, England, the Pont de Blois, France, and, the Krämerbrücke, Erfurt, Germany.

NEWCASTLE

In terms of the number and variety of buildings erected on bridges in England during the Middle Ages, the three most important examples were those across the River Tyne at Newcastle, across the River Avon at Bristol and across the River Thames in London. The inhabited bridge at Newcastle was destroyed by floods in 1771, and the other two were demolished and replaced in 1764 and 1831 respectively.

Old Tyne Bridge provided a critical crossing point on the main thoroughfare from the east of Scotland to the South (fig. 31). It stood on the site of a Roman bridge constructed by Hadrian. Ownership of the medieval bridge was shared between the ecclesiastical and municipal authorities, a situation which could provoke considerable friction. When, for example, the mayor of Newcastle erected a tower on the third of the bridge owned by the Bishop of Durham, the Bishop entred into legal proceedings and, with judgement in his favour, took possession of the tower and reasserted his ownership of the Church's portion of the bridge. A blue stone on the bridge marked the division between the north and south of the county palatine and between the two proprietors of the bridge. On the southern side, urban development was dense with houses rising four storeys on both sides of the street. On the southern side, towards Gateshead, the building density was confined to the sites immediately above the bridge's piers (fig. 32). Damaged beyond repair in 1771 by floods, the medieval bridge was reconstructed on a scale two-thirds of the original as part of the Royal Jubilee Exhibition of 1887.

Fig. 32, John Hilbert, *Newcastle Bridge,* 1727, engraving. Newcastle City Libraries and Arts

BLOIS

The most informative visual record of the medieval Pont de Blois is the plan and elevation drawn by Poictevin on 12 February 1716 (fig. 33). Reconstructed from what remained of the bridge after its destruction by ice on 6-7 February 1716 and from Poictevin's recollections, the drawing shows the bridge to have possessed all the attributes of an urban inhabited bridge. As part of the city's defensive system, it contained the Porte Saint-Fiacre, a massive square gateway and portcullis dating from the eleventh century, other defensive towers and two further drawbridges. At least five watermills found space upon the bridge, and it catered for the spiritual well-being of its users by the construction, probably in the eleventh century, of a chapel dedicated to Saint Fiacre. From the twelfth century the bridge was believed to have been entirely covered by commercial and residential buildings, their development encouraged by Louis XII who accorded the municipality the privilege of building in return for annuities. Although most of the commercial and residential buildings had been destroyed by the time of Poictevin's drawing, the bridge had provided accommodation for such tradesmen as butchers, cobblers, rope-makers, potters and coopers. It suffered much damage and repair over the centuries, including the demolition of some of its arches by the citizens of Blois to block the advance of the Huguenots in the 1560s. It also witnessed the passage of almost every king of France as well as that of Joan of Arc in 1428.

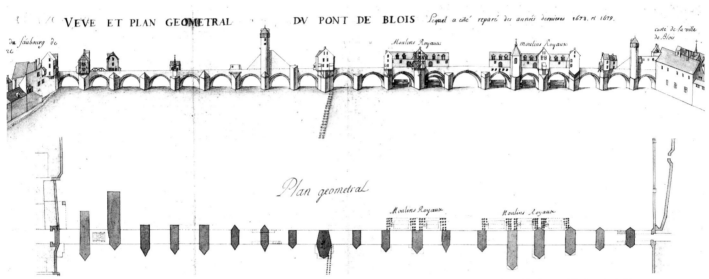

Fig. 33, Nicholas Poictevin, *Plan and view of the bridge at Blois as repaired in 1678-79*, 1680, pen and brown ink with wash. Bibliothèque Nationale de France

ERFURT

The Krämerbrücke at Erfurt, with buildings along the entire stretch of its bridging structure, is one of the rare examples of a surviving medieval bridge. Built in the tenth to eleventh centuries initially as a wooden structure spanning the River Gera between the Wenigemarkt, and the fish market and town hall, it was by the following century to accommodate small kiosks and shops. It was reworked in stone and completed in 1235 from funds generated by the buildings erected on it (fig. 34). It subsequently added a church at each end and, during the Middle Ages, possibly a convent. Other examples of inhabited bridges which acquired similar multiple use existed throughout Europe, ranging from the more single functional mill-bridges with residential units above, as at Meaux, France (fig. 35), destroyed by fire on 17 June 1920, to more composite structures such as the bridge at Esslingen (fig. 36).

Fig. 34, Erfurt, *Krämerbrücke: View of the 14th-century bridge over the River Gera*, photograph. © Remy Bouguennec 1996

Fig. 35, Meaux, *Pont du Marché*

Fig. 36, Esslingen am Neckar, *Innere Brücke*, mid-19th century, lithograph. Stadtarchiv Esslingen am Neckar

Fig. 37, Avignon, *Pont Saint-Bénézet*, 19th century, lithograph. Bibliothèque des Arts Décoratifs, Paris, Collection Maciet

CHAPEL BRIDGES

The partially ruined bridge of St-Bénézet at Avignon is one of the few surviving examples of a bridge with a chapel on it (figs. 37, 38). Such edifices were, however, fairly common in the Middle Ages (see also figs. 39, 40) when a bridge defying the raging fury of the waters flowing beneath it was viewed with a certain reverence. The construction of a chapel, usually dedicated to St Nicholas, patron saint of sailors, was the concrete manifestation of such reverence and was often financed by donations, alms or even papal or episcopal indulgences. Such bridges appear to have originated in Italy around the mid-twelfth century (fig. 39), but spread rapidly northwards to the rest of Europe (fig. 40).

According to legend, the bridge at Avignon was built by Saint Bénézet, a shepherd who had been ordered to construct a bridge by an angel and who, together with his disciples, collected the necessary funds and assured its realization in 1177-85. A wooden bridge was indeed built on this site at about that date. It was replaced by a stone structure after its destruction in 1226. Originally 850 metres long, the bridge linked Avignon to Villeneuve, crossing the island of Barthlasse and spanning the two arms of the River Rhône. Its arches were frequently destroyed by floods, most notably in the early seventeenth century: only four of the original twenty-two arches still stand.

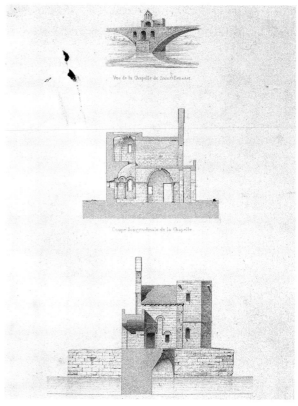

Fig. 38, Avignon, *Pont Saint-Bénézet: three details*, lithograph. Bibliothèque des Arts Décoratifs, Paris, Collection Maciet

The chapel is Romanesque in style, with an upper floor having been inserted during the thirteenth century. In 1513, a gothic apse was added to the upperpart of the structure, together with an adjoining lodging. Mass was last celebrated in the chapel for the last time in 1715.

Chapel bridges survive in France, at Pernes-les-Fontaines and Bar-le-Duc, in Germany at Calw, and in the United Kingdon at Bradford-on-Avon, Wakefield and Rotherham, Yorkshire (now over land) and St Ives (Huntingdonshire).

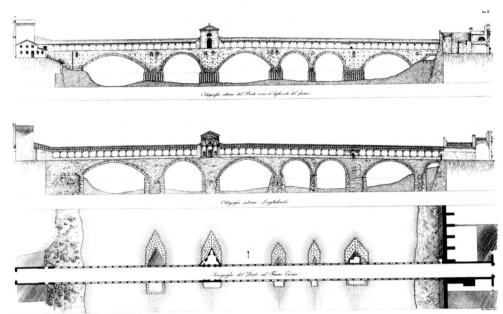

Fig. 39,
Pavia, bridge over the Ticino River:
elevation and cross-section,
from Giovanni Voghera,
Monumenti pavesi,
Pavia 1828. Civica Raccolta
delle Stampe A. Bertarelli,
Castello Sforzesco, Milan

Fig. 40, Hans Leu, *View of Zurich with the Wasserkirche,* 1500, oil on panel. Baugeschichtliches Archiv der Stadt Zürich

FORTIFIED BRIDGES

On account of their majestic appearance, fortified bridges continued to inspire bridge-builders long after they had lost their defensive role. Examples include the bridge at Chatellerault in France, constructed early in the seventeenth century and Tower Bridge in London dating from the late nineteenth century.

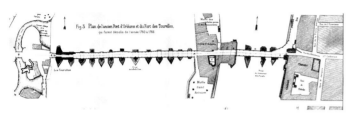

Fig. 41, *Orléans, plan of the bridge in 1423 before the siege by the English,* Plate II, *Atlas des Mémoires de la Société Archéologique et Historique d'Orléans.* Archives Départementales, Orléans

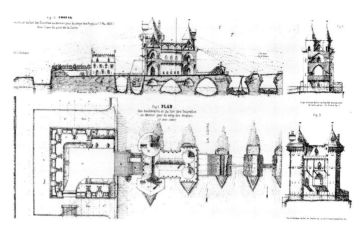

Fig. 42, *Orléans, the Fort des Tourelles on the bridge: elevations and plans,* Fig I, M.A. Collin, *Le Pont des Tourelles à Orléans,* Orléans 1895. Archives Départementales, Orléans

Traditionally, however, fortified bridges were an integral part of the defensive system of cities adjacent to rivers. With the changing character of military tactics and the expansion of cities beyond their walls, the fortified bridge lost its original function. Many fortified towers on bridges disappeared after the eighteenth century since their presence contricted the flow of the traffic .

The element most commonly found on a fortified bridge was the defensive tower with portcullis and shutters which blocked the enemy's advance. In some cases, the bridges had drawbridges or removable sections of the roadway as a further deterrent. In others, the system of fortification included more complex structures. At Beaugency, for example, the four arches closest to the town were surmounted with walls and flanked by towers upon the nosings of the piers, while at Orléans the bridge had two drawbridges, a fortified portal, a bastille and, at one end, the many-towered Fort des Tourelles; the bridge was demolished between 1760 and 1766 (figs. 41, 42).

Pont Valentré at Cahors, built over the Lot between 1308 and 1380, remains intact (fig. 43). 170 metres long and spanning a 127 metre-wide expanse of water, it consists of six large gothic arches and two smaller ones and supports three towers of approximately 40 metres high. The rectangular median tower probably served as an observation post, whereas the two square lateral towers clearly had a specifically defensive role, since they are fitted with machicolations and cross-shaped openings for archers. It successfully defied the English during the Hundred Years War and Henri de Navarre (the future Henri IV of France) during the Siege of Cahors in 1580.

In England, the medieval bridges at Bath, Newcastle (see fig. 31) and Durham had defensive towers and Bristol Bridge had a strong defensive gate; Monmouth, in Wales, possesses an extant example, originally constructed in 1272, which consists of a forbidding tower and gatehouse. Other surviving examples of fortified bridges can be seen at Orthez, Sauveterre-de-Béarn and Sospelin France; Finstermünz in Switzerland; and at Besalú and Frias in Spain.

Fig. 43,
Joseph Southall,
Pont Valentré de Cahors, 1936, watercolour.
Birmingham Museums and Art Gallery

RURAL BRIDGES

While this catalogue concerns itself primarily with urban inhabited bridges, their rural counterpart demands investigation. Usually created within the strictly private domain, the rural inhabited bridge tended to be the product of personal fantasy. It can be classified into two broad categories: château-bridges, which included residential accommodation, and garden bridges, which were primarily decorative; the Blenheim project (see p. 45) falls somewhere between the two. France possesses two château-bridges: the Château de Fère-en-Tardenois, now in ruins (fig. 44), and the Château de Chenonceaux in the Loire Valley (figs. 45, 46).

In 1526, François I gave Anne de Montmorency, Governor-General of France, a castle at Fère, lying to the east of Paris between Château-Thierry and Soissons. Thirteen years later, the architect Jean Bullant was asked to restore this fortress-like castle, perched upon a hill overlooking a precipice. He replaced the castle's fortified drawbridge with a grandiose three-metre-wide Italianate bridge supported on four piers spanning a 55 metre-wide chasm. Above rose a double arcade, the lower containing a passageway and the upper, reached by a staircase, an observation gallery. The new structure was completed in the late 1550s.

Fig. 44, *Château de Fère-en-Tardenois in 1775*, engaved by R Peltier, 1855. Château de Fère-en-Tardenois

43

The history of the inhabited bridge at the Château of Chenonceaux is more complicated. In the early sixteenth century, Thomas Bohier, chamberlain to Charles VIII, constructed a château on the right bank of the River Cher. It included a four metre wide arcade linking the entrance gate to two full-length windows with balconies overlooking the river on the ground and first floors. In 1547, Henri II, then owner of the castle, gave it to his mistress Diane de Poitiers who appointed the architect Philibert de l'Orme in 1556. He planned a bridge across the river, surmounted by an arcade, which would lie immediately adjacent to the earlier château in order to leave unimpeded the view from Bohier's full-length windows. This project was not completed, for, on the death of Henri II in 1559, Diane de Poitiers was obliged to leave Chenonceaux. Its new proprietor, Catherine de Médicis, ordered the bridge's completion, but in a modified form so that it was connected to the left bank of the river merely by a drawbridge. Only on de l'Orme's death in 1570, when she appointed Jean Bullant as architect, was the building finally completed, with three floors running the full length of the bridge to the same height as the castle.

Fig. 46,
View of the Château de Chenonceaux, with the completed gallery, 17th century, engraving. Conservateur du Château de Chenonceaux

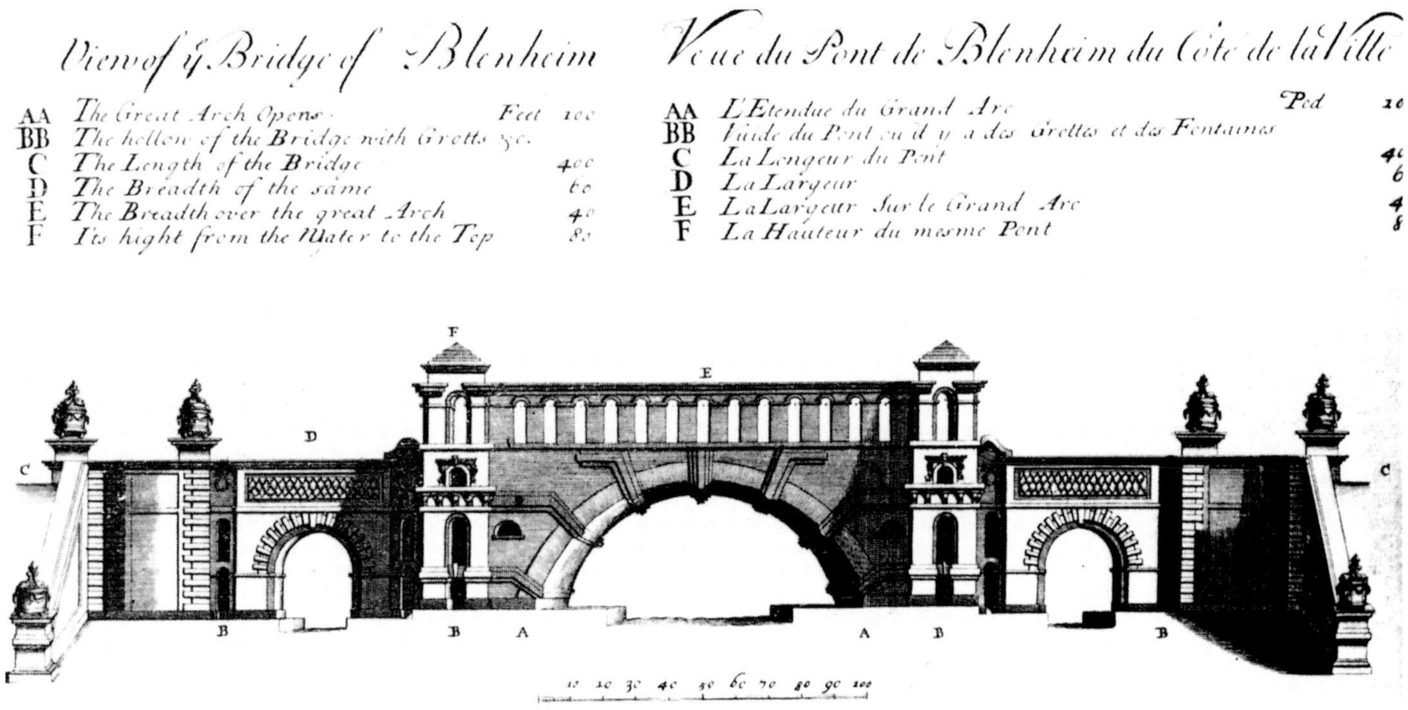

Fig. 47, Sir John Vanburgh, *View of the Bridge of Blenheim*, 1705, in Colen Campbell, *Vitruvius Britannicus*, vol I, 1717. Royal Academy of Arts

The rural inhabited bridge intended for the park at Blenheim Palace, England, could have represented a cross between a structure destined for domestic use and a garden folly on a monumental scale (fig. 47). In 1705, John Churchill, 1st Duke of Marlborough, received from Queen Anne in gratitude for his defeat of Louis XIV the Royal Manor of Woodstock and monies with which to build a mansion there. He entrusted the creation of a magnificent palace and garden to the architects John Vanbrugh and his assistant Nicholas Hawksmoor. As part of the landscaping of the park, they designed a monumental bridge to carry the Grand Avenue across the River Glyme to the house. Work began in 1708. The bridge, 31 metres in length, was designed to support a 24 metre high arcaded building containing as many as thirty-three rooms. The Duchess of Marlborough, however, disapproving of this extravagant 'bridge in the air', cancelled the project in 1712 and the arcade was never completed. The lower rooms, which had been finished by that date, were flooded when 'Capability' Brown transformed that stretch of the River Glyme into an ornamental lake in 1764.

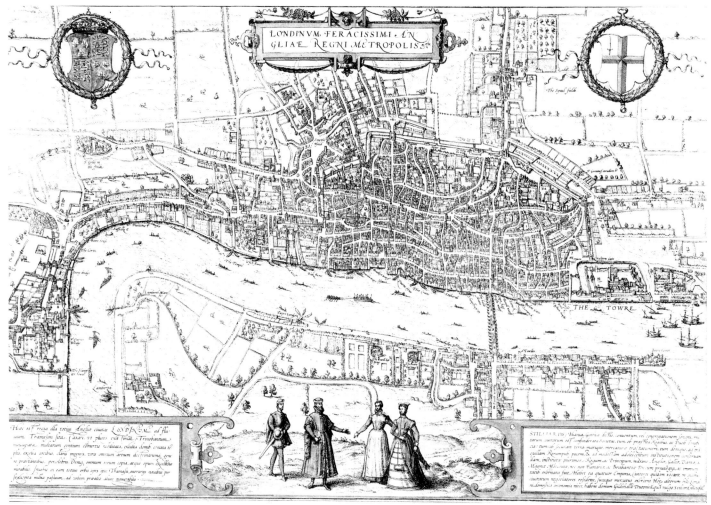

Fig. 48, Frans Hogenberg, *Londinium Feracissimi Angliae Regni Metropolis,* engraving, Georg Braun and Frans Hogenberg, *Civitas Orbis Terrarum,* 1572. © Guildhall Library, Corporation of London

London

OLD LONDON BRIDGE

Old London Bridge represents a prime example of a multifunctional inhabited bridge, combining commercial buildings, domestic dwellings – often on a considerable scale – a chapel, industrial structures and warehouses. It had a life span of over six hundred years before it was finally demolished in 1823.

There had been wooden bridges on the site since the time of the Roman occupation but the first stone bridge was built between 1176 and 1209 by Peter, chaplain of St Mary Colechurch. It crossed the River Thames in London between Southwark and the City. Until 1739, it was the only bridge in London (figs. 48, 49). The bridge consisted of a stone platform, erected on elm piles driven into the river

bed, and is recorded to have measured 285 metres long, 4.6 metres wide and 18.5 metres high, with twenty arches (fig. 50). The first record of houses on the bridge dates from 1201. Above the central and longest pier a chapel dedicated to St Thomas à Becket was erected and was reputed to have been very elegant with an upper and a lower chamber 18.5 metres high. It was converted into a warehouse and dwelling in 1737.

The history of the buildings on the bridge is complex owing to the various disasters afflicting the bridge itself (fig. 50). During much of the thirteenth century, it was in a reputedly piteous state and it suffered particularly in 1212-13, when fire broke out, and in 1282, when five of the arches were swept away during a great frost, but it was repaired in 1300. It is known to have had a tower on it in the early thirteenth century but this was soon to be destroyed by fire. It also had a drawbridge, and a second tower was built at its northern end in 1426. The most famous building on the bridge was Nonesuch House; set over the seventh and eighth arches from the Southwark end, it had been prefabricated out of wood in Holland, shipped to London and erected on the bridge using wooden pegs. Less spectacular, timber-framed houses lined a street along almost the whole length of the bridge (fig. 51). In 1582, a Dutchman, Peter Morris, built the first waterworks in order to supply the city with water and three other mills were built six years later for grinding corn. In a fire of 1632-3, more than forty houses on the bridge were destroyed but their replacements of 1645-6 were also destined for a similar fate owing to the Great Fire of London of 1666 (fig. 54). During the subsequent five years, a new range of four-storey houses was erected and the causeway extended from 5 metres to 6 metres in width; the houses on the southern side were also rebuilt during this time in order to give the bridge a greater architectural uniformity (fig. 55).

Old London Bridge exerted a powerful influence on the pattern of London's urban exapnsion. As the city's sole crossing point of the River Thames, the bridge tended to encourage high density development on its northern bank. At its southern end it established lighter patterns of building as they fanned out along the river bank and into Southwark.

The bridge also played an important symbolic and political role in the life of the city. Until fire struck the bridge once again in 1725, traitors' heads were displayed at the Southwark end, and it took central stage for such major historical events as Charles II's progress to reclaim his throne at the Restoration of the Stuart monarchy in 1660.

The bridge underwent further alterations during the eighteenth century. Following reports drawn up by Labeyle, the engineer of Westminster Bridge, then under construction, and Dance, the city surveyor, a temporary wooden bridge was erected on the west side, the houses were demolished, the central pier and two adjoining arches were removed to create one great arch, and the bridge itself widened. Complaints about the bridge's instability, its insecurity and the dangers for navigation continued, however, and were acompanied by remarks about its unpleasing appearance in comparison to the new bridges of the time. In 1823-31, it was finally replaced by a new bridge of five stone arches, built upstream by Sir John Rennie to a design by his late father.

Fig. 49, Anglo-Dutch School, *London from Southwark*, c. 1630, oil on panel. Museum of London, © Museum of London

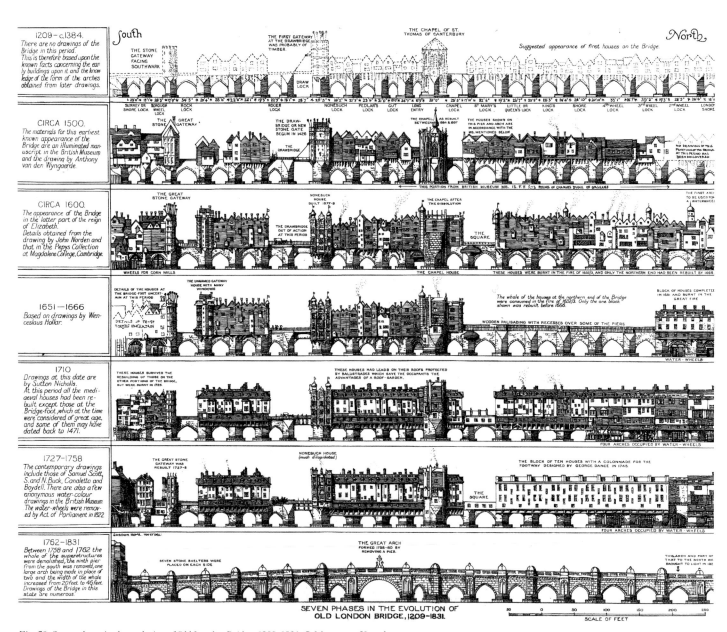

Fig. 50, *Seven phases in the evolution of Old London Bridge, 1209-1831.* © Museum of London

Fig. 51, Claude de Jongh, *Old London Bridge*, 1630, oil on panel. Kenwood House, London

Fig. 52 Abraham Hondius, *The Frozen Thames looking westwards towards Old London Bridge*, 1677, oil on canvas. Museum of London

Fig. 53, Anon, *A Frost Fair on the Thames at Temple Stairs*, c. 1684, oil on canvas. Museum of London

Fig. 54, Dutch School, *The Great Fire of London*, 1666, c. 1666, oil on panel. Museum of London

Fig. 55, Antonio Joli, *The Thames from Somerset House*, c. 1747, oil on canvas. Natwest Group Art Collection

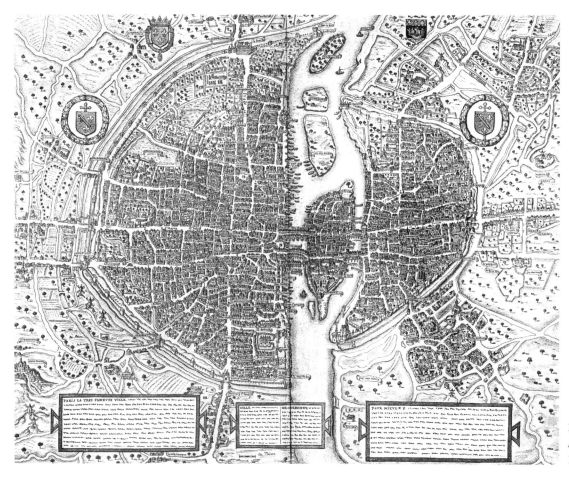

Fig. 56,
Map of Paris around 1540.
Paris, Musée Carnavalet

Paris

The city of Paris has always boasted a large number of bridges (over thirty lie within the Boulevard Periphérique today) and it undoubtedly had the highest number of inhabited bridges of any European city between the twelfth and the eighteenth centuries. Such a density can be explained in both geographic and urbanist terms. The four principal bridges – Pont au Change, Pont Notre-Dame, Pont Saint-Michel and the Petit Pont – permitted access between the Île de la Cité in the centre and the left and right banks of the river at a pivotal point between the Roman north/south land axis and the east/west navigation axis. Geographically, the width of the river and the position of the island were ideal for such bridges, the river being neither so wide as to split the city (as in London or Budapest) nor so narrow, as in Amsterdam, that it was not worthwhile to cover or inhabit the bridges. In terms of its urban development, Paris spread out in concentric circles from the Île de la Cité. Thus the river and island retained their focal position, in contradistinction to cities such as London where one bank initially developed far more rapidly than the other. The need to contain the city within defensive walls encouraged a pattern of dense urban development and the choice by the royal family to reside at the western end of the fortified city during the early centuries encouraged schemes for the city's architectural embellishment (figs. 56, 57).

Commercial stalls are known to have been set up on the Parisian bridges as early as the twelfth century and by the beginning of the fifteenth century the Pont Notre-Dame boasted systematically built uniform rows of shops and houses. This latter bridge and its replacement of the early sixteenth century (see below) can be considered the archetype of the bridge-street tradition which was to be emulated in other Parisian bridges and which privileged the interior aspect of the bridge (fig. 57). Quite frequently the internal façades of these bridges underwent new decoration (temporary or permanent), particularly when they served as the venue for celebrating events such as victories, peace treaties, royal births, coronations or marriages. Unlike, for example, the projects designed for the Ponte di Rialto in Venice which favoured the monumental view of the bridge from the river (see pp. 66-71), the bridges of Paris were to be admired on the inner street sides often virtually ignoring the presence of the river (fig. 59).

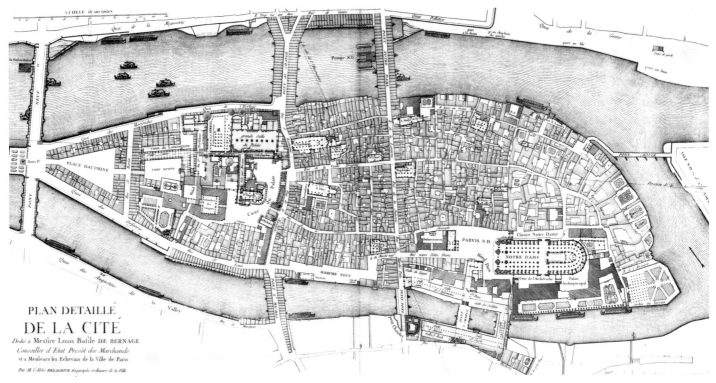

Fig. 57, Abbé Delagrive, *Detailed Map of the City, showing the Île de la Cité*, 1754, engraving. Bibliothèque Nationale de France

This may have been the reason that ensured that the Pont Neuf, begun in 1578, supported no superstructures save a pumphouse and embrasures for kiosks placed above the piers. An inhabited bridge in the manner of the Pont Notre-Dame would have presented the monarch with a view from the Louvre encumbered by a disorganized array of housebacks, windows, balconies and lavatories.

The history of Paris's inhabited bridges is extremely complex as they underwent many transformations including changes of name,

destruction by flood, damage by frost or fire, pressure of traffic and reconstruction. They were initially built of wood, gradually becoming rendered in stone. Paris never enjoyed an inhabited bridge constructed of iron, despite Gustave Eiffel's dramatic scheme for the Pont d'Iéna in 1878 (see fig. 133). By the time the city's first iron bridge, the Pont des Arts, was constructed, the inhabited bridge had become a discredited building type on the grounds of sanitation, security and aesthetics.

Fig. 58, *Paris, the Plan Turgot (detail)*, 1734. Paris, Musée Carnavalet

Fig. 59, Balthazar Probst, *The Pont Saint-Michel viewed from the Pont Neuf*, 17th century. Paris, Musée Carnavalet

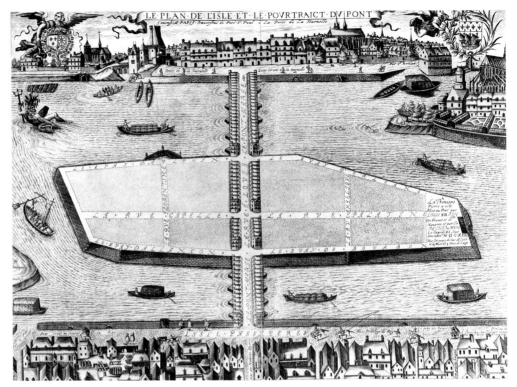

Fig. 60,
J. Messager and J. Siveline,
*Plan of the Île Saint-Louis and the view
of the bridge under construction in
Paris, crossing from Port Saint-Paul
to the Porte de la Tourelle*,
1614, engraving.
Paris, Musée Carnavalet

THE PONT MARIE AND THE ÎLE SAINT-LOUIS

Although the Île de la Cité had been urbanized since the origins of Paris, the Île Saint-Louis (previously known as the Île Notre-Dame) did not become built-up until the seventeenth century. Its development was conditional upon the creation of two inhabited bridges, or, as they were both to be called the Pont Marie, one bridge in two segments linked by a road, the rue des Deux Ponts (fig. 60). This would not only ensure the island's connection to the mainland on each side but would also be seen to guarantee the island's rapid and successful urbanization. In 1614, an entrepreneur, Christophe Marie, drew up an agreement with Nicolas Bruart, Chancellor of France, acting on behalf of the king, giving him permission to build upon the Île Saint-Louis provided that he undertook to construct, within ten

years, two inhabited stone bridges as clearly depicted in Siveline's engraving of the time. Although Louis XIII and the Queen Mother, Marie de Médicis, laid the first stone in the same year, the bridges were never built according to the original design of a double row of symmetrical houses similar to those on the Pont Notre-Dame. Only one of the two halves of the bridge was finally completed with its houses in 1643 (fig. 61) but their life was short since swollen river water destroyed two arches and twenty-two houses in 1658. The houses were not replaced and the remaining ones were partially destroyed in 1741 following flood damage and finally demolished in 1786. The other section of the bridge, later to be renamed the Pont de la Tournelle, was never to be inhabited.

Fig. 61, Nicolas and Jean-Baptiste Raguenet, *The Pont Marie and the Île Saint-Louis*, 1757, oil on canvas. Paris, Musée Carnavalet

Fig. 62, Nicolas and Jean-Baptiste Raguenet, *The joust of the watermen between the Pont Notre-Dame and the Pont au Change (looking towards the Pont au Change)*, 1756, oil on canvas. Paris, Musée Carnavalet

THE PONT AU CHANGE

The Pont au Change is the name of the bridge built across the Seine from the Right Bank to the Île de la Cité from 1639 to replace the Grand Pont (fig. 62). It was designed by du Cerceau (see pp. 60, 61). A royal patent letter issued that year instructed that, in keeping with the contemporary architectural taste, the buildings on the bridge should all be constructed from the same materials and to the same height. Two rows of uniform houses featured shops at ground level opening on to the central street (fig. 63) with closed balconies on the river façades, kitchens on the river side on a mezzanine above the shops, three upper floors and an attic on the fourth floor. An edict of 1786 ordered the demolition of the houses and this was carried out during the succeeding two years, in sympathy with the

Fig. 63, *View of the Pont au Change, begun in 1639 during the reign of Louis XIII and completed under Louis XIV*, undated, etched by Aveline. Paris, Musée Carnavalet

Teste vers la rue de la barrillierie du pont a bastir a Paris

CE Pont doit estre basty à la place, où estoit anciennement en l'Année 1338. & long-temps auparauant le Pont lors vulgairement nommé le Grand Pont de Paris, où depuis a esté le Pont au Change, où estoient les Orpheures & le Pont Meusnier où y auoit quantité de Moulins à blé, en la place duquel Pont Meusnier, apres sa cheute, à esté basty le Pont aux Oyseaux.

Fig. 64,
Marcel Le Roy
*Design for a bridge to be built in Paris
on the site of the Pont au Change:
entrance to the bridge*, 1622,
engraved c. 1639.
Bibliothèque Nationale de France

opinion expressed by Moreau in his urban plan of the 1760s: 'The river's channel, once opened up, will offer the most extensive and wonderful vista that one could find in a large city' ('le canal de la rivière entièrement libre, offrira le spectacle le plus vaste et le plus magnifique qu'on puisse trouver dans une grande ville').

The name Pont au Change refers back to the functions of previous bridges on this site, most particularly those on the Grand Pont built in 1141, upon which Louis VII had ordered that money changers were to ply their trade exclusively within the city. The history of previous bridges on the site is complex, not least because, when the Grand Pont collapsed in 1296, its replacement was positioned very slightly upstream in order for the King to avoid financial obligations to an ecclesiastical fiefdom. Subsequent bridges, all with shops along

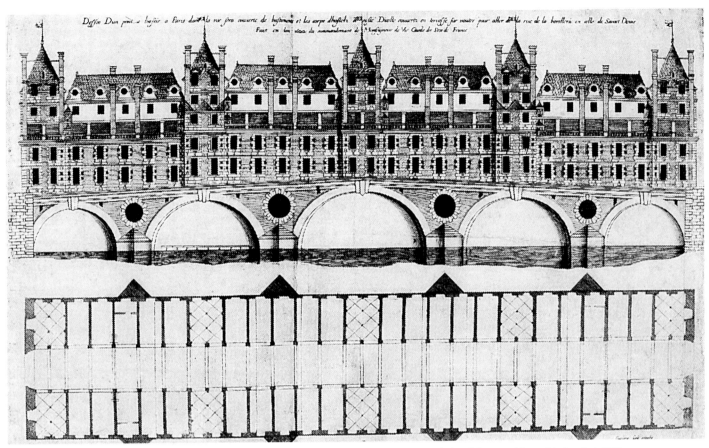

Fig. 65, Marcel Le Roy, *Design for a bridge to be built in Paris on the site of the Pont au Change: elevation and plan of the ground floor of the first deck*, 1622, engraved c. 1639. Bibliothèque Nationale de France

their causeways, included the Grand Pont aux Changeurs (1298-1621), the Pont des Meuniers (1323-1596) and the Pont Marchant (1609-1621) which was also built according to a strictly geometrical scheme dictated by the king. The outcome of this repositioning was that the Pont au Change of 1639 stood upon the sites of both the Grand Pont of 1141 and its 'neighbour', the Pont Marchant, thus becoming a rare example of a Y-shaped bridge (see figs. 57, 63). Probably the most interesting proposal for the construction of a new bridge on the site was that made by Marcel Le Roy in 1622 but unrealized (figs. 64, 65). It is possible that this scheme represents the first shopping mall bridge, whose street was lit by oil lamps and whose superstructure was sufficiently flexible in plan to allow for shops with houses above and an open terrace on one floor which could serve as a promenade for viewing the river.

Fig. 66,
(Anne Fonbonne)
Visiting card for Theodore de Hansy's bookshop on the Pont au Change,
1768, engraving.
Paris, Musée Carnavalet

Fig. 67, *Perspectival view of the Pont Notre-Dame built in 1507 under the direction of Jean Jucundus Cordelier, native of Verona,*
no date, colour engraving by Aveline. Paris, Musée Carnavalet

THE PONT NOTRE-DAME

The Pont Notre-Dame linked the Right Bank of the River Seine to the Île de la Cité upstream from the Pont au Change. Originally constructed as an inhabited bridge in 1414-1419, it was succeeded by one constructed between 1500 and 1512 (see fig. 58).

Both bridges played an important role in the life of the city. The fifteenth-century bridge had boasted an early example of a rectilinear street with a regular succession of uniform elements on the façades and a triumphal arch at the end of the cornices. The essentially sceniographic nature of its architecture ensured it a role as a triumphal bridge: national celebrations included processions through its portals and porticoes (fig. 69).

The collapse of the houses on the bridge in 1499 killing four or five people apparently caused an outcry in the city and a number of people responsible for its maintenance were punished by life imprisonment. De Felin, Maître des oeuvres de la Ville, was charged with the planning and building of the second bridge and is believed to have enlisted the collaboration of Fra Giocondo (fig. 70). The stone bridge, 124 metres long and 24 metres wide, supported two rows of thirty-four houses reached through shops at ground level (figs. 70, 71). The shops were apparently brighter than usual, lit by an uninterrupted sequence of large windows of equal size (also permitting a wider display of goods) framed by the arches of an innovative portico. Further light was later provided through windows fitted on the river side.

A variety of goods was sold on the bridge, including jewellery, paintings, weapons, clothing, some foodstuffs, pharmaceutical articles and perfume. The introduction of the mass production of glass saw an increased number of dealers in mirrors and chandeliers. The art gallery owned by Gersaint has been immortalized in the painting *Gersaint's Shopsign* by Antoine Watteau (fig. 68) who lived andworked with the dealer for several months.

58

Fig. 68,
Jean-Antoine Watteau,
Gersaint's Shopsign,
1720, oil on canvas.
Schloss Charlottenburg,
Staatliche Schlösser und
Gärten, Berlin

Fig. 69,
Anon,
*Triumphal arch raised at
the end of the Pont
Notre-Dame*, no date,
engraving. Paris,
Musée Carnavalet

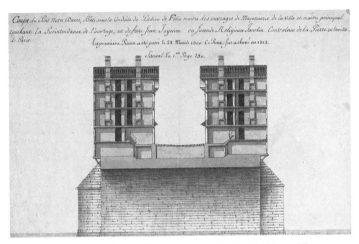

Fig. 70, *Cross-section of the Pont Notre-Dame attributed to de Felin and Fra Giocondo*, Reproduced in D. M. Federici, *Convito Borgiano*, 1792. Bibliotecea Civica, Treviso

Like its predecessor, this handsome bridge was frequently employed for official ceremonies, starting with the entry of the wife of François I in 1531. In 1660, on the occasion of the festivities for the triumphal entry of Louis XIV and Marie-Thèrése, the bridge was redecorated as shown in the engraving by Aveline (fig. 67). The walls boasted sculpted terms carrying baskets of flowers and fruit linked together by garlands and medallions representing the kings of France accompanied by inscriptions; the ceremonial effect was completed by the creation of a triumphal arch by the Beaubrun brothers.

In the 1760s it was decided to demolish the buildings on the bridge. However, as they were a source of revenue, this was delayed until 1786 (see fig. 76). Once stripped of its massive superstructures the bridge conformed to the 'rational' thinking of the day and was temporarily rebaptized 'the bridge of reason' during the French Revolution.

Fig. 71, *Elevation of the Pont Notre-Dame attributed to de Felin and Fra Giocondo,* reproduced in D. M. Federici, *Convito Borgiano*, 1792. Biblioteca Civica, Treviso

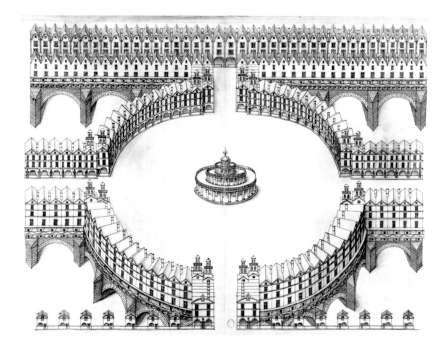

Fig. 72,
Jacques I Androuet du
Cerceau, *Project for the Pont Neuf,*
c. 1578, ink on vellum.
Bibliothèque Nationale de France

THE PONT NEUF

This remarkable project by Jacques I Androuet du Cerceau, architect to Henry III, was designed about 1578 for a site now held by the Pont Neuf at the western tip of the Île de la Cité (fig. 72). A new bridge was required to alleviate the pressure of traffic on the Pont au Change and the Pont Notre-Dame.

Rather than being a very precise proposal for that specific site, this drawing should be considered as a model which resumes the urban thinking of the architect and his age. Whereas two centuries later the Enlightenment was to dismiss the idea of an inhabited bridge, the bridge street toward the end of the sixteenth century was considered the most noble form of bridge with which to adorn a great city. The Pont Notre-Dame was reputedly du Cerveau's point of reference and he included it, repositioned, in his drawing. However, du Cerceau takes his concept further by associating the bridge street with the new notion of the bridge square. The potential of the combination of these two urban models could then be developed throughout the city, transforming it with a sequence of such bridges into a grandiose metropolis on water, reminiscent of Venice or Bruges.

The bridge that was actually built across the two arms of the River Seine was reputedly designed by du Cerceau and Des Illes.

Fig. 73, Anon, *Unexecuted project for the Pont Neuf,* before 1578, oil on canvas. Paris, Musée Carnavalet

Fig. 74, Nicolas and Jean-Baptiste Raguenet, *The Pont Neuf, showing the Samaritaine*, 1755, oil on canvas. Musée Nissim de Camondo, Paris

Fig. 75, *The Pont Neuf: one of the last kiosks*, 1847, lithograph. Bibliothèque des Arts Décoratifs, Paris, Collection Maciet

Du Cerceau's son, Baptiste, who frequently collaborated with his father on projects and was Surindentant des Bâtiments, may also have produced a design for a bridge on the same site. The drawings have been lost, but the proposal was for an inhabited rather than merely vehicular structure. The first stone of the Pont Neuf was laid by Henri III in 1578 but political events culminating in his assassination in 1589 interrupted construction. The bridge was finally completed under Henri IV in 1606. Its pavement was a novelty, as was its classical appearance, unencumbered by any building (fig. 73).

A painting of 1755 by Nicolas and Jean-Baptiste Raguenet depicts the Pont Neuf and the building known as the Samaritaine which housed a water pump (fig. 74) erected in 1608 by the Flemish engineer Lintlaer. The tradition of colonizing a bridge with structures, albeit makeshift, died hard, and the painting shows Parisians establishing a makeshift 'inhabited bridge' with market stalls. These were given a brief spell of permanence when small stone kiosks were constructed above the piers of the bridge (fig. 75). These were demolished in the 1850s.

Fig. 76, Hubert Robert, *Demolition of the houses on the Pont au Change*, 1786-87, oil on canvas. Staatliche Kunsthalle, Karlsruhe

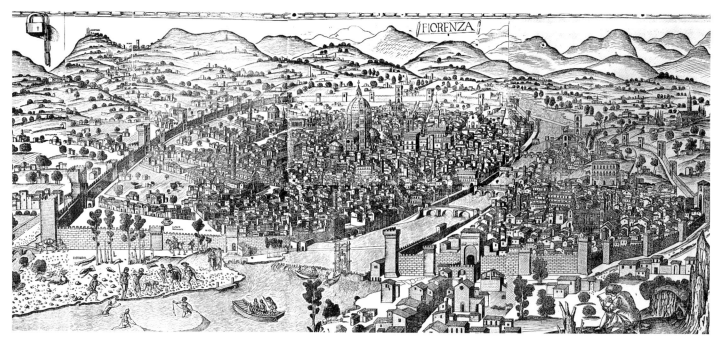

Fig. 77, Lucantonio degli Uberti, *Florence: the Chain Plan*, 1470, wood engraving. Kupferstichkabinett, Staatliche Museen zu Berlin

Florence

PONTE VECCHIO

The Ponte Vecchio ('Old Bridge') is the third bridge on its site crossing the River Arno (fig. 77). At the time of its completion it was known as the 'new' old bridge. Its predecessor the 'Ponte Vecchio – so-called from 1218 in order to distinguish it from the Ponte de la Carraia downstream on the River Arno at S. Trinità and the Ponte Rubaconte upstream – had already replaced the original wooden structure, which had collapsed after the flood of 1177, with wooden buildings supported by five stone arches. This 'Ponte Vecchio' fulfilled a central role within the urban layout of Florence after the new town walls had been built in 1172 to encompass the bank on the far side of the Arno (fig. 78). It had supported a church, the homes and towers of the Mannelli family (ancient guardians of the bridge), a vegetable market and, by the early fourteenth century, forty-three shops. Although this bridge had supported a particularly dense number of buildings and variety of functions, it was not exceptional for its time. The neighbouring Ponte alle Grazie, for example, which survived until the twentieth century, sported a roadway flanked by houses, hermitages, oratories and a small church.

The new Ponte Vecchio of 1345 belonged to the Commune; both the bridge and its forty-seven shops were constructed out of stone in a single building campaign (fig. 79). Organized in four groups with an empty space in the centre of the span, the shops were originally occupied by a variety of tradesmen, from butchers and grocers to blacksmiths.

However, in 1593, the Grand Duke Ferdinand I de' Medici decreed that these should be replaced by luxury trades such as gold- and silversmithery and money-changing. The latter decree was a direct result of the substantial transformation that the bridge had recently undergone as part of the extensive programme launched by Ferdinand's father Cosimo I de' Medici in order to redesign the area around the Palazzo dei Signori and its link to the new Palazzo Pitti across the River Arno. Under the direction of Giorgio Vasari, a new 'corridor' was constructed in 1565 above the existing buildings, providing the Prince and his Court with an exclusive route from the façade of the new Uffizi overlooking the River Arno across the buildings along one side of the Ponte Vecchio (which entailed the construction of three arches to support its weight in the central section) and on to the Palazzo Pitti. From 1593, the shops continued to offer luxury goods and they were redecorated and additions were built over the river (fig. 80). In the mid-nineteenth century, as part of a major urban redevelopment plan for Florence, it was proposed that the Ponte Vecchio should be remodelled with a glass and iron superstructure. Designed by Martelli and Corazzi, it was never executed (see fig. 130).

Fig. 78,
Hendrik van Cleve,
View of Florence,
1583, pen and ink,
with ink wash.
Istituto Nazionale
per la Grafica, Rome

Fig. 79, Bernardo Bellotto, *View of the River Arno looking towards the Ponte Vecchio*, c. 1742, oil on canvas. Beit Collection, Russborough

Fig. 80, Israel Silvestre, *View of the Ponte Vecchio, seen from the Uffizi in Florence*, mid-17th century, pencil with grey, green and pink wash.
The Metropolitan Museum of Art, Rogers Fund 1963

Fig. 81,
Anon, *View on the Ponte Vecchio,
with the Ponte alle Grazie
in the background* (detail),
c. 1850, oil on canvas.
Collection Grassi

Fig. 82,
View across the Ponte Vecchio,
1898, photograph.
Archivi Alinari, Florence

Fig. 83,
The Ponte Vecchio,
photograph.
© C. M. Bednarski 1996

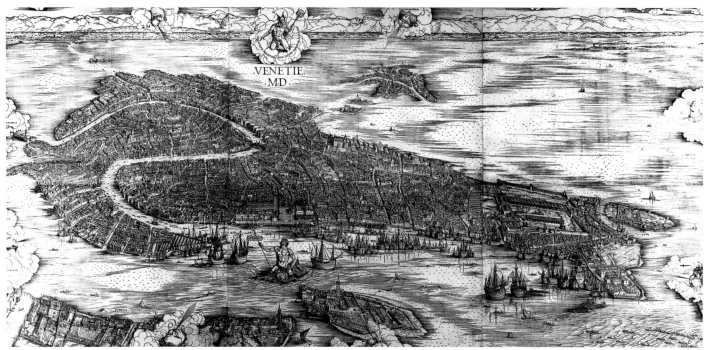

Fig. 84, Attributed to Jacopo de' Barbari, *Bird's-eye View of Venice*, 1500, woodcut from six blocks. Department of Prints and Drawings, British Museum

Venice

PONTE DI RIALTO

The Ponte di Rialto, built between 1588 and 1591, is possibly the most famous of all inhabited bridges. Its history is closely related to that of its predecessor on the same site and, at the time of its constuction, the fervour aroused among architects and public alike was sparked not by the presence of shops along its causeway but by the fact that it had no piers: the two islands of Rivoalto and Luprio were audaciously linked across the Grand Canal in a single span (see fig. 91).

A wooden bridge (figs. 84, 85) had crossed the Grand Canal on the site of the Ponte di Rialto since 1250. It had a central drawbridge which allowed ships to pass beneath and could separate, if required, the two islands. Situated in a vital position as the unique link by foot to the eastern island and its increasingly important market, traffic on the bridge grew rapidly, attracting in its wake the unregulated presence of squatters and salesmen. In an attempt to control this development, the magistracy in charge of the Rialto, the salt

purveyors (Ufficiali sopra Rialto, Provveditori al sale), approved the construction of two rows of shops along the bridge in the first half of the fifteenth century, the rent providing income which could contribute to the maintenance of the bridge.

Considerable concern over the vulnerability of the bridge, especially to fire, led to a proposal in 1503 to replace the wooden bridge with a stone structure. Debate continued sporadically throughout much of the sixteenth century and projects were advanced by architects as expert as Fra Giocondo (also involved in the Pont Notre-Dame in Paris (see pp. 58, 59). A consultation process, launched by the Senate in 1551, led to a number of proposals by architects including Sansovino, Vignola, Gugliemo di Grande (fig. 86), Marastoni (fig. 87) and particularly Palladio, who produced various drawings for the site from the 1550s. None of these designs managed to inspire sufficient interest owing mainly, it would appear, to their utilization of several structural arches and/or classically inspired designs

Fig. 85, Vittore Carpaccio, *The Miracle of the True Cross*, c. 1494, oil on canvas. Galleria dell'Accademia, Venice

Fig. 86, Francesco Lazzari, after Gugliemo di Grande, *Design for a stone bridge*, original design, 1587, reconstructed 1880, pencil, pen and ink. Museo Civico Correr, Venice

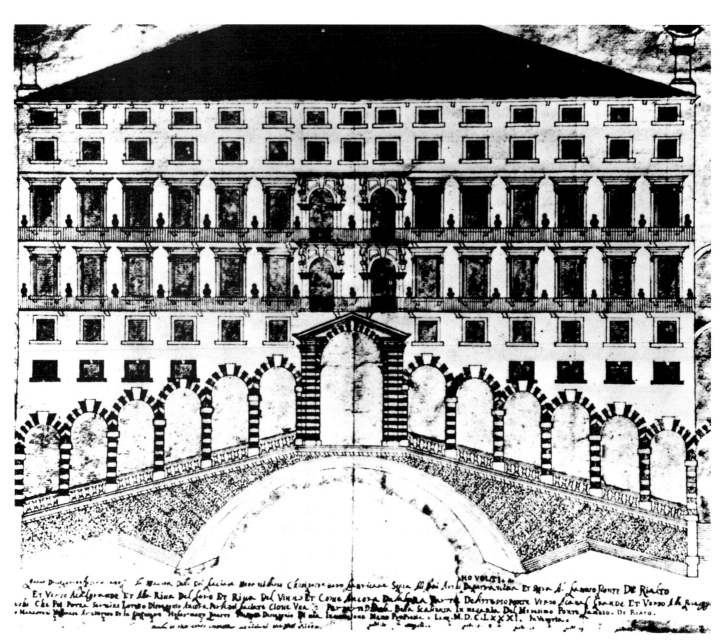

Fig. 87, Gugliemo Marastoni, *Design for the Ponte di Rialto*, 1580, pen and ink. Venice, Archivio di Stato

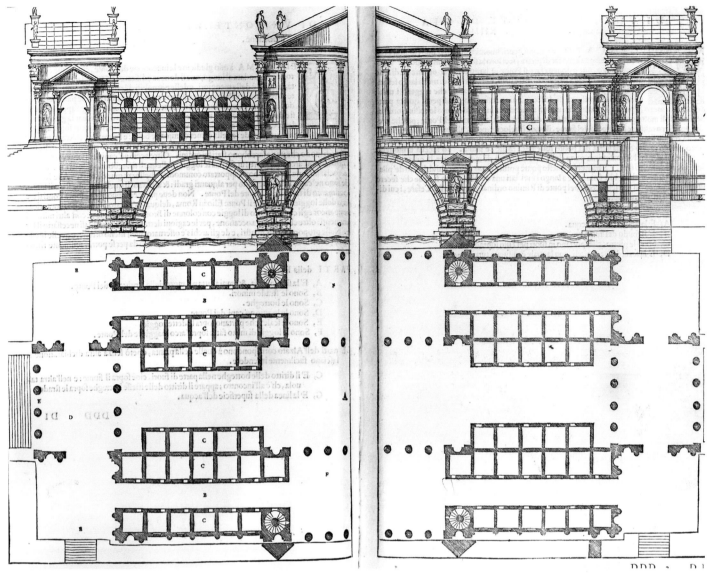

Fig. 88, Andrea Palladio, *Design for the Ponte di Rialto*, 1554, published, *I Quattro Libri dall'Architettura*, Book III. British Architectural Library, RIBA, London

Fig. 89,
Canaletto,
*Venice: cappriccio view
with Palladio's design for
the Ponte di Rialto,*
1742, oil on canvas.
The Royal Collection

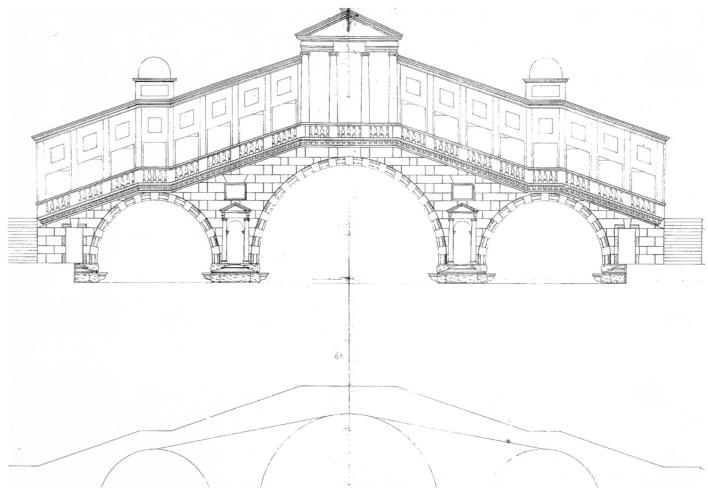

Fig. 90, Vincenzo Scamozzi, *Design for the Ponte di Rialto*, 1588, pen and ink. British Architectural Library, RIBA, London

considered more appropriate to Rome than to Venice. In Palladio's earliest design, the bridge rested on five arches and supported a central dome, but he paid virtually no attention to the detail concerning the row of shops. However, he envisaged the bridge as the central element in an intentionally provocative urban plan which replaced several existing buildings by a forum at either end of the bridge. These fora gradually disappeared in later reworkings of his proposal and in 1570 he published an engraving in *I Quattro Libri* which depicts the later stages of his thinking (fig. 88). Palladio's schemes were to inspire artists and architects alike. Both Canaletto (fig. 89) and Guardi painted interpretations of his bridge, the former placing it in one version within an imaginary landscape of the master's buildings (see *Cappriccio: Ponte Palladiano*, 1755, Galleria Nazionale, Parma), and architects designed Palladian-style bridges (see pp. 74, 75).

The increasingly heated debate over the preferred form of the bridge continued until, on 7 January 1588, the Senate finally adopted the option of a single-arched span. The senators entrusted the responsibility of building the new bridge to Antonio dal Ponte, the architect favoured by the salt purveyors. While expressing no specific design preference, the senators did lay down certain guidelines: the course of the bridge's axis; a central footway between two rows of shops, and two side paths protected by low balustrades permitting unobstructed views of the Grand Canal. These guidelines appear to have been based on a design for the bridge contained in a drawing which is believed to have been by Vincenzo Scamozzi and now resides in the collection of the Royal Institute of British Architects (fig. 90). Scamozzi had added a text to his drawing dated 1 January 1588 in which he violently but unsuccessfully opposed the idea of a single arch. Even though this drawing differs quite considerably from Antonio dal Ponte's executed design, certain features – such as the balustrade and the central opening in the rows of shops – were incorporated.

Despite a widespread climate of disbelief about the technological viability of its audacious, single-span design, the bridge was completed in 1591 and quickly acquired an international renown which it has never lost (figs. 91, 92), even inspiring an imitation in Nuremberg.

Fig. 91, Canaletto, *The Ponte di Rialto*, oil on canvas. Musée du Louvre, Paris

Fig. 92, Francesco Guardi, *The Ponte di Rialto*, oil on canvas. Musée des Augustins, Toulouse

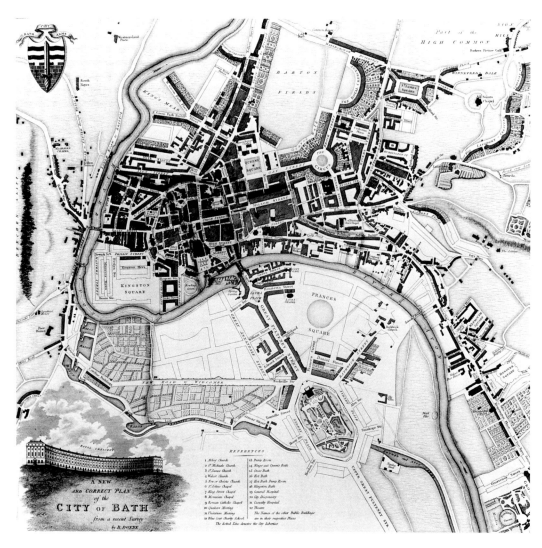

Bath

PULTENEY BRIDGE

Pulteney Bridge, spanning the River Avon in Bath, is one of the last inhabited bridges to have been built. Its construction coincided somewhat ironically with the period of destruction of the majority of inhabited bridges in many other European towns and cities (see fig. 76).

The fashionable Georgian spa town of Bath, laid out by John Wood the Younger, lay on the north bank of the River Avon (fig. 93). In 1767 Frances Pulteney, wife of the Edinburgh lawyer William Johnstone Pulteney, inherited the 600-acre estate of Bathwick. The lawyer immediately recognized the estate's development potential for an expansion of Bath on the river's southern side. However, to guarantee such a result, the existing river crossing, provided by a ferry,

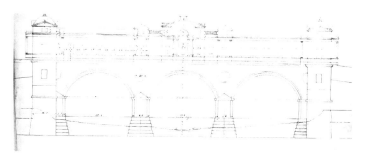

Fig. 94, Robert Adam, *Design for Pulteney Bridge, Bath*, c. 1770.
By courtesy of the Trustees of Sir John Soane's Museum, London

Fig. 95, Thomas Malton, *Pulteney Bridge, Bath, from the river*, 1785, watercolour.
Victoria Art Gallery, Bath and North-East Somerset

would have to be replaced by a bridge. In 1768 the initial scheme for a simple vehicular bridge had been drawn up. Two years later, however, Pulteney turned to Robert Adam, and the scheme became more grandiose. During his Grand Tour, Adam had visited both Florence and Venice. He was also familiar with Palladio's unrealized proposals for the Ponte di Rialto (see fig. 88). His design for Pulteney Bridge (fig. 94) gave him an opportunity to make a similarly dramatic architectural statement, as well as to provide commercial facilities on the bridge for generating income. The bridge was completed in 1773 (fig. 95). It was supported by two piers and three arches, above which rose two rows of eleven small shops with attics above and, in some cases, cellars below, on either side of the roadway (fig. 96). It is probable that the north and south façades were originally of identical design. However, shortly after Adam's death in 1792, Pulteney employed the architect of the Bathwick estate, Thomas Baldwin, to modify the buildings, most notably to raise the height of the first-floor rooms. Following serious damage inflicted upon the structure by floods and storms in 1799 and 1800, the north side was rebuilt by Baldwin's successor, John Pinch. Over the course of the twentieth century, Bath City Council has progressively reinstated Adam's original façade (fig. 97).

Fig. 96, Thomas Malton, *Pulteney Bridge, Bath from Bridge Street,* 1777, pencil and watercolour. Victoria Art Gallery, Bath and
North-East Somerset Council

Fig. 97, Pulteney Bridge, Bath

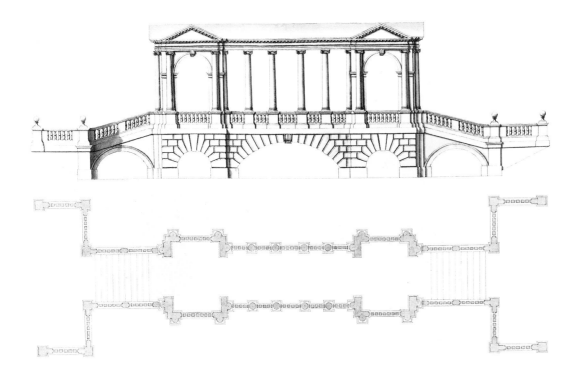

Fig. 98,
Plan and elevation of the Earl of Pembroke's Bridge at Wilton in Wiltshire, in Colen Campbell, *Vitruvius Britannicus*, ed., 1771

The Palladian Bridge

England possesses three major examples of a 'Palladian bridge': at Wilton, near Salisbury (fig. 98); at Prior Park near Bath (fig. 100); and at Stowe, in Buckinghamshire. Drawing their inspiration from designs published in Palladio's *I Quattro Libri*, these bridges, despite their superstructures, performed primarily an ornamental function within the context of a landscaped park. Their creation reflected the British architectural style dominant during the eighteenth century, namely neo-Palladianism (fig. 99). Other examples are thought to have existed at Dogmersfield Park, Hampshire, and at South Lodge, Middlesex.

In 1736-37, in the landscaped grounds of Wilton House, the Architect Earl, Henry Herbert, 9th Earl of Pembroke, in collaboration with Roger Morris and the sculptor John Devall, created an ornamental pedestrian bridge (fig. 101). Its superstructure consisted of an Ionic colonnade of four columns and two half-columns linking two pavilions. Pembroke had access both to Palladio's original drawings, then in the collection of Lord Burlington, and to *I Quattro Libri*. The bridge at Wilton was subsequently engraved in Volume V of Woolf and Gandon's 1771 edition of *Vitruvius Britannicus* (fig. 98).

Fig. 99, Robert Adam, *Design for a bridge in a Palladian style*, chalk.
By courtesy of the Trustees of Sir John Soane's Museum, London

Fig. 100, Thomas Robin, *Prior Park, Bath: The Palladian Bridge*,
No 24, from Thomas Robin's sketchbook, 1762,
pencil, pen and ink wash. Private Collection

James Gibbs proposed a Palladian bridge for the grounds at Stowe in about 1738. Designed to carry vehicles, it had a blind wall with a bas-relief by Scheemakers and frescos by Francesco Sleter. The blind wall was subsequently replaced by a colonnade designed by Giambattista Borra in 1762. Palladio's original designs for a bridge, as well as the example at Wilton, may have encouraged Ralph Allen to commission the builder Richard Jones to construct a Palladian bridge at Prior Park, near Bath, in 1755.

Outside England, a Palladian bridge was built in 1774 to designs by V. Neelov for Catherine the Great at her summer palace of Tsarkoye Selo, outside St Petersburg.

Fig. 101, Antonio Visentini and Francesco Zuccarelli, *The Bridge at Wilton*, 1746, oil on canvas. Private Collection, Venice

Fig. 102, Giovanni Battista Piranesi, *A Bridge of Magnificence with Loggias and Arches built by a Roman Emperor*, published in G B Piranesi, *Prima Parte di Architetture e Prospetti*, 1743, etching. Bibliothèque des Arts Décoratifs, Paris, Collection Maciet

The Triumphal Bridge

The taste for triumphal bridges coincided with the decline of the inhabited bridge in Europe. It lasted from the late eighteenth to the early twentieth century and was almost exclusively academic and visionary in intent. Its sources lay in the imaginary reconstructions of ancient architecture, as expounded for example in J. B. Fischer von Erlach's history of architecture published in 1721, and epitomized in Giovanni Battista Piranesi's etched architectural compositions published from 1743 (figs. 102, 103).

The theme of the triumphal bridge was employed as an exercise for architectural students in the Académie Royale d'Architecture, Paris, from 1774 and subsequently at the École des Beaux-Arts throughout the nineteenth century. Characterized by an abundance of triumphal arches and colonnades, the compositions were grandiose in intent (figs. 105-107) and rarely bore any relation to the realities of contemporary bridge-building (fig. 104). Indeed, during the late eighteenth century with the rise of the professional engineer – to whom was increasingly entrusted the design of new bridges unencumbered by architectural superstructures and with larger spans – the designs for triumphal bridges established a rapprochement between architects and artists, whose compositions celebrated these imaginary constructions (fig. 119).

Fig. 103, Ennemond-Alexandre Petitot,
Project for a triumphal bridge, 1748, engraving.
Bibliothèque des Arts Décoratifs, Paris, Collection Maciet

Fig. 104, Davey de Chavigne and Gustave Taraval, *Pont de la Liberté*, 1781,
engraving. Bibliothèque des Arts Décoratifs, Paris, Collection Maciet
The triumphal bridge was designed to link the eastern end of the Île de la Cité
to the Île Saint-Louis, Paris.

Fig. 105,
J. B. L. F. Lefebvre,
*A Triumphal Bridge: elevation,
project presented to the Ecole
Royale des Beaux-Arts*, 1786.
Ecole Nationale Superieure des
Beaux-Arts, Paris

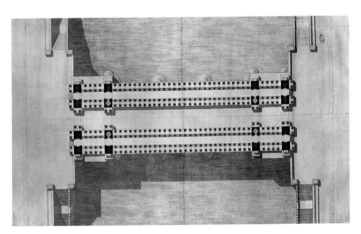

Fig.106,
J. B. L. F. Lefebvre, *A Triumphal Bridge:
plan, project presented to the Ecole Royale
des Beaux-Arts*, 1786. Ecole Nationale
Superieure des Beaux-Arts, Paris

Fig. 107,
J. B. L. F. Lefebvre,
*A Triumphal Bridge:
Cross-section, project presented
to the Ecole Royale
des Beaux-Arts*, 1786.
Ecole Nationale Superieure des
Beaux-Arts, Paris

Fig. 108,
Eduard Gärtner,
*The Königsbrücke, with the
Königskolonnaden, Berlin*,
1835, oil on canvas. Berlin,
Stadtmuseum Berlin

Fascination for the triumphal bridge spread throughout Europe, inspiring schemes which were both realized, as in the case of the Königsbrücke (1762), Berlin (fig. 108), and megalomaniacal, as in the bridge proposed by Otto Wagner as the gateway to his pantheon of art in 1880 (fig. 109). In England, as early as 1759, Sir William Chambers had entered a competition for a bridge to be constructed at Blackfriars, in London, with a design which consisted of a columnar superstructure placed over the central arch (fig. 110). Likewise, as at the Ecole Royale d'Architecture, the subject lent itself to competition projects set by the Royal Academy of Arts from its foundation in 1768. Thomas Sandby, first Professor of Architecture at the Royal Academy, designed a 'Bridge of Magnificence' in a late neo-Palladian style around 1770 (figs. 111, 112). Unrolled at his sixth architecture lecture delivered in 1774, the design for a 357 metre-long bridge was

Fig. 109, Otto Wagner, *A Pantheon of Art: aerial perspectival view of the Pantheon built on an artificial lake surrounded by colonnades and triumphal arches*, 1880, pen and ink. Historisches Museum der Stadt Wien

Fig. 110, William Chambers, *Competition Design for Blackfriars Bridge*, engraved from a drawing in the British Architectural Library. British Architectural Library, RIBA, London

Fig. 111,

Thomas Sandby, *Design for a Bridge of Magnificence:*
perspective of the interior looking towards the domed terminal pavilions,
c. 1770, pen and ochre wash.
British Architectural Library, RIBA, London

Fig. 112, Thomas Sandby, *Design for a Bridge of Magnificence: elevation*, 1780. By courtesy of the Trustees of Sir John Soane's Museum, London

Fig. 113, Joseph Gandy after John Soane, *Perspectival drawing of a Triumphal Bridge*, 1776/1799, watercolour.
By courtesy of the Trustees of Sir John Soane's Museum, London

Fig. 114, Joseph Gandy after John Soane, *Design for a Triumphal Bridge: elevation*, 1776/1799. By courtesy of the Trustees of Sir John Soane's Museum. London

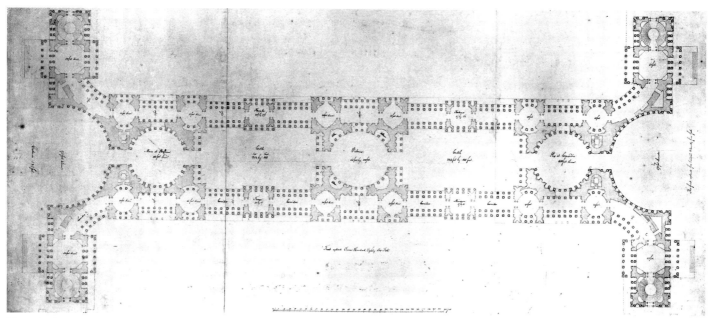

Fig. 115, John Soane, *Plan of the superstructure of a Triumphal Bridge*, 1776. By courtesy of the Trustees of Sir John Soane's Museum. London

Fig. 116, Joseph Gandy after John Soane, *Design for a Triumphal Bridge*, 1799, watercolour. By courtesy of the Trustees of Sir John Soane's Museum. London

Fig. 117, Joseph Gandy after John Soane, *Design for a Triumphal Bridge (in the Corinthian style)*, 1799, watercolour. By courtesy of the Trustees of Sir John Soane's Museum, London

Fig. 118, Joseph Gandy after John Soane, *Design for a Triumphal Bridge (in the Doric style)* 1799, watercolour. By courtesy of the Trustees of Sir John Soane's Museum, London

reported to have been a dramatic event and to have created a powerful impression upon John Soane, then a student in the Royal Academy Schools. The design was subsequently exhibited at the Royal Academy's annual exhibition of 1781.

Sandby's design, initially for an unspecified site in London but later proposed for one close to where Waterloo Bridge now stands, was soon surpassed in the scale of its ambition by the 365 metre-long design entered by John Soane for the Royal Academy Gold Medal in 1776 (figs. 113, 114). Soane's scheme consisted of a central, domed

rotunda with two open rotundas at either end, linked by a continuous colonnade surmounting a seven-arched structure. The bridge was anchored to the banks by domed buildings entered across semicircular columnar forecourts. Soane returned to this project later in his life, reworking the Corinthian columns into the more severe Doric style, in keeping with a contemporary shift in taste (fig. 118). Joseph Gandy's perspective drawings of Soane's designs capture the sublime nature of this visionary scheme, the spirit of which is also translated into paint in J.M.W. Turner's *Ancient Rome. Agrippina landing with the Ashes of Germanicus* (1839; fig. 119).

Fig. 119, J. M. W. Turner, *Ancient Rome, Agrippina landing with the Ashes of Germanicus*, 1839, oil on canvas. Trustees of the Tate Gallery, London

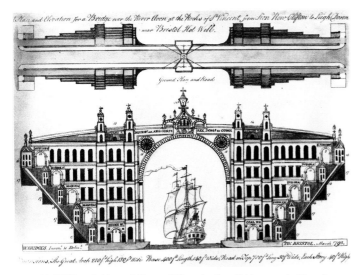

Fig. 120, William Bridges, *A Plan and Elevation for a Bridge over the River Avon at the Rocks of St Vincent's, from Sion Row Clifton to Leigh Down near Bristol Hot Wells*, 1793, engraving. British Architectural Library, RIBA, London

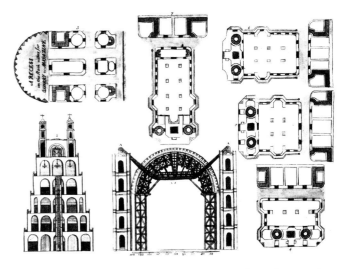

Fig.121, William Bridges, *A bridge over the River Avon: plans and cross-sections*, 1793, engraving. British Architectural Library, RIBA, London

Bristol

AVON GORGE BRIDGE

Virtually nothing is known about William Bridges, the author of a spectacular design made in 1793 for an inhabited bridge outside Bristol to cross the Avon Gorge from Sion Row, Clifton, to Leigh Down, near Bristol Hotwells.

The need for a bridge at this point had long been recognized. As early as 1753, William Vick, a wine merchant, had bequeathed £1,000 to the Society of Merchant Venturers of Bristol to accumulate at compound interest until it had reached £10,000, when it was to be used to build a toll-free stone bridge across the gorge at Clifton. It was not until 1829-36 that Isambard Kingdom Brunel finally spanned the gorge with his masterpiece, the Clifton Suspension Bridge.

Although William Bridges' project might provoke derision, it was nevertheless intended to be a serious proposal, since its author had his drawings engraved by P. Daniel and accompanied by a printed proposal (figs. 120, 121). His financial arguments prefigure those put forward by, for example, the American architects in the 1920s (see pp. 94-96), for he claimed that the cost of construction could be redeemed by rent gathered from its supporting buildings.

Unlike earlier inhabited bridges where buildings stood astride the road deck, Bridges' design envisaged a huge five-storey substructure supporting the roadway and a huge central arch, measuring some 71 metres in height and some 60 metres in width, to permit the passage of vessels from Avonmouth to the Bristol docks. The five storeys, each 12 metres high, in the abutments of the bridge were designed to fulfill a variety of functions. The proposed rooms included public granaries, a corn exchange, wharves and storage for coal, a general market, a museum and library, a marine school, offices, stables, warehouses and twenty dwellings. Above the arch would have been a chapel, a toll house (contrary to Vick's express desire), and a belfry with a lighthouse at its apex. In its spandrels there would have been two windmills.

The Nineteenth Century

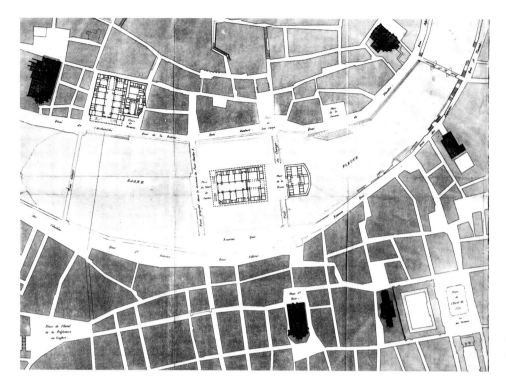

Fig. 122,
L. P. Baltard Lyon: *Map of the central part of the city, showing the proposed location for the law courts and the prison over the River Saône*, 1828. Archives Municipales, Lyon

BALTARD'S PROPOSAL FOR LYON

Despite the demolition of most inhabited bridges in Europe by the end of the eighteenth century, the building type as a viable design solution continued to fascinate architects. Several schemes produced over the succeeding two centuries were the result of unsolicited proposals for specific sites made by their creators.

The bridge proposed by L. P. Baltard for Lyon would have replaced an existing bridge – the Pont du Change over the River Saône – which had incorporated during its lifetime a sequence of superstructures. Built in the eleventh and twelfth centuries, the Pont du Change linked 'episcopal' Lyon on the right bank to the 'municipal' city on the 'presqu'île' (fig. 122). It lay across a bed of granite rocks, known as the 'rochers du Change', into which were sunk several of its piers. In the 1560s, drawings and even a model are believed to have been prepared for a new town hall to be built upon these rocks. In the seventeenth century, homes and shops were built on part of the bridge and in the early nineteenth century a café, the 'Neptune', was erected there.

In 1827-28, a limited competition was organized to design a new courthouse and prison for Lyon to replace the outmoded 'Palais de Roanne' on the river bank. As no particular site was specified, Baltard suggested two alternatives: that of the existing court house, and that of the Pont du Change and its 'rochers'. His project was

Fig. 123, L. P. Baltard, *Project for a theatre and shops to be built on the Pont au Change: elevation*, c. 1828, lithograph. Bibliothèque Municipale de Lyon, Coste 666

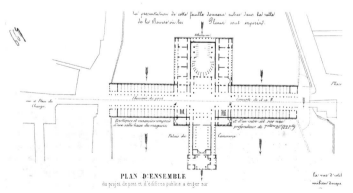

Fig. 124, L. P. Baltard, *Project for a theatre and shops to be built on the Pont au Change: plan*, c. 1828, lithograph. Bibliothèque Municipale de Lyon, Coste 667

Royal Academy of Arts

LIVING BRIDGES

26 September – 18 December 1996. Open 10am – 6pm daily

EXHIBITION

Journey through the Royal Academy along a river of time and discover how inhabited bridges, real and imagined, can change our cities

COMPETITION

Thames Water Habitable Bridge Competition. Seven international architects exhibit their visions of a new bridge for London

Supported by and GENERALE DES EAUX GROUP   in association with THE INDEPENDENT   Supported by

Twenty-one scale models, beginning with Old London Bridge, the Ponte di Rialto in Venice and Pont de Notre-Dame in Paris, will span a real river transforming the Main Galleries of the Royal Academy. 'Living Bridges' will look at the impact that inhabited bridges - with shops, businesses and homes - have had on our cities. The exhibition looks at inhabited bridges from Medieval times to the present day and includes a dramatic crystal Tower Bridge as well as designs for bridges in London by such celebrated architects as Richard Rodgers and Terry Farrell. The 'river of time' will be complemented by photographs, models and projections.

Royal Academy of Arts
Burlington House
Piccadilly
London W1V 0DS
0171- 439 7438

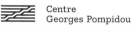

Centre
Georges Pompidou

Exhibition realised with the collaboration of the Centre Georges Pompidou, Musée national d'art moderne, Paris

John Norden 'View of London Bridge' (detail) c. 1600
© Museum of London

LIVING BRIDGES

Thames Water Habitable Bridge Competition
To coincide with 'Living Bridges', seven international architectural practices have been invited to design a new inhabited bridge crossing the River Thames from the London Television Centre on the South Bank to Temple Gardens on the North Bank. Models of all seven entries, including the winner, will be shown at the exhibition. Visitors will have their chance to vote on the proposals and the most popular public design will be announced towards the end of the show.

26 September-18 December 1996
Open 10am-6pm daily
Last admission 5.30pm

Tickets
Full price   £5.00
Concessions   £3.50
12-18 years   £2.50
8-11 years   £1.00

Tickets may be bought in advance, by post (please include 50p p+p) or in person from the Royal Academy Ticket Office or by telephone on 0171-494 5676. Ticket office and telephone bookings open 10am-5.30pm, daily.

Save Money on a Thames River Trip
Present your 'Living Bridges' ticket stub at either Tower Pier or Westminster Pier and save £1.00 per ticket on a City Cruises river trip. See the site of the proposed Thames Water Habitable Bridge. Phone 0171-930 9033 for details.

The Academy Card
The Academy Card at £15.00 entitles you to visit five different exhibitions of your choice over an 18 month period. For further details call 0171-494 5664.

Free mailing list
Please call 0171-494 5614 to join.

How to get to the Royal Academy
Tube to Green Park or Piccadilly Circus or buses 9, 14, 19, 22 and 38

ROYAL ACADEMY OF ARTS PRESS RELEASE

PICCADILLY LONDON W1V 0DS TEL: 0171 494 5615 FAX: 0171 439 4998
REGISTERED CHARITY NUMBER 212798

THAMES WATER HABITABLE BRIDGE COMPETITION
in conjunction with the exhibition
LIVING BRIDGES
26 September - 18 December 1996

TWO WINNERS IN DESIGN COMPETITION
FOR A NEW "OLD LONDON BRIDGE"

The Secretary of State for the Environment, The Rt. Hon. John Gummer MP, announced that **Antoine Grumbach & Associates** of France and **Zaha M. Hadid** of the UK are the joint winners of the Thames Water Habitable Bridge Competition to design a new inhabited bridge across the River Thames. The proposed site of the bridge is from Temple Gardens to the area in front of the London Television Centre on the South Bank.

The two winners were selected from a shortlist of seven internationally renowned architectural practices which included **Architectburo Libeskind** (Germany), **Branson Coates** (UK), **Future Systems** (UK), **Ian Ritchie Architects** (UK), and **Krier Kohl** (Germany).

Professional vs public opinion
Models and drawings of the seven proposals will form the centrepiece of the exhibition **Living Bridges** which looks at the history of the inhabited bridge. There will be an opportunity to discover whether public opinion coincides with the decision of the assessors. All visitors to the exhibition **Living Bridges** will be encouraged to vote for their own favourite design. An announcement of the public's choice will be made on 9 December.

The assessors were Sir Philip Dowson PRA (Chairman), Rt. Hon. John Gummer MP, Michael Cassidy (Corporation of London), Sir Robert Clarke (Thames Water), Jean Dethier (Centre Georges Pompidou), Gordon Graham PPRIBA, Peter Murray (exhibition curator) and Janet Street-Porter (broadcaster).

Living Bridges (supported by the Corporation of London and the Générale des Eaux Group in association with The Independent) is at the Royal Academy of Arts from 26 September to 18 December 1996. The exhibition has been developed from a concept initiated by, and has drawn on research undertaken by the Centre Georges Pompidou, Musée national d'art moderne, Paris, under the direction of Jean Dethier.

For further information or photographs please contact Katharine Jones, Press and Promotions Officer, or Linda Weston, Deputy Press Officer on 0171-494 5615/5610 or fax on 0171-439 4998.

25.9.96

ROYAL ACADEMY OF ARTS PRESS RELEASE

PICCADILLY LONDON W1V 0DS TEL: 0171 494 5615 FAX: 0171 439 4998
REGISTERED CHARITY NUMBER 212798

LIVING BRIDGES   **MAIN GALLERIES**

The inhabited bridge, past, present and future

26 September - 18 December 1996

Supported by and GENERALE DES EAUX GROUP in association with THE INDEPENDENT

realised with the Centre Georges Pompidou
Musée national d'art moderne. Paris

Exciting proposals for a new inhabited bridge for the River Thames in London will form the centrepiece of the exhibition *Living Bridges*. *Living Bridges* illustrates the history and possibilities of inhabited river crossings, from Old London Bridge with its assortment of shops and houses to new proposals for similar structures in cities around the world. Variously built as fortifications, dams, toll-gates, mills, marketplaces, even hospitals or convents, inhabited bridges have caught the imagination of architects throughout the ages.

An innovative design by the architect Nigel Coates will create an eye-catching installation at the Royal Academy: a spectacular river will flow through the Main Galleries at Burlington House, spanned by twenty-one specially commissioned models of bridges, both real and imagined, designed over the last 600 years. The first model is of Old London Bridge as it appeared in 1600, and the sequence continues chronologically and includes the Ponte Vecchio in Florence and the Pont de Notre-Dame in Paris (one of four inhabited bridges in that city). Other celebrated examples, such as the Ponte di Rialto in Venice and Pulteney Bridge, Bath, the last inhabited bridge to be built in Britain, in 1770, are included. Models of proposed bridges which never left the drawing board have been created especially for the exhibition, such as Sir John Soane's idea for a triumphal crossing of the River Thames.

Models of inhabited bridges of the 20th century include a re-creation of Sir Edwin Lutyens' plan for an art gallery bridge across the River Liffey in Dublin, a daring 1940s scheme to replace Tower Bridge with a glass tower, as well as a 1960s proposal by Jellicoe and Coleridge for a Crystal Span bridge for the River Thames at Vauxhall. Contemporary proposals for London will include Lifschutz Davidson's recent winning design for Hungerford Bridge, Terry Farrell's Blackfriars Thameslink 2000 and two of the Peabody Trust's inhabited bridge schemes of 1995 by Richard Horden and Allies & Morrison, between St Paul's and Bankside. The exhibition's climax will present London as it could be: the results of the **Thames Water Habitable Bridge Competition** will be on display and will conjure a vision of the River Thames for the 21st century.

A COLLABORATION BETWEEN THE ROYAL ACADEMY OF ARTS, LONDON AND THE CENTRE GEORGES POMPIDOU, PARIS

Living Bridges is the first major collaboration between the two international arts organisations, the Royal Academy of Arts and the Centre Georges Pompidou. The exhibition was initiated by Jean Dethier, exhibition director and architectural advisor at the Centre Georges Pompidou, in cooperation with Ruth Eaton, who has acted as historical curator. It has been developed for presentation at the Royal Academy by Peter Murray, who was founder-publisher of Blueprint magazine, and is now Principal of Wordsearch Communications, and by MaryAnne Stevens, Education Secretary and Chief Curator at the Royal Academy with specific responsibility for architectural programmes.

COMPETITION

A competition, supported by Thames Water Plc, was launched by the Royal Academy in April, to design a new inhabited bridge which will span the river from Temple Gardens on the North bank to the the area in front of London Television Centre on the South. Designs by Antoine Grumbach & Associates and by Zaha Hadid Architects were chosen as joint winners by the competition assessors. Please see separate release for further details.

PUBLIC VOTE

All seven entries in the competition will be on display in the exhibition **Living Bridges**, and the public will be given the opportunity to vote for their favourite design. The public's choice will be announced on 9 December, and will reveal whether public opinion coincides with that of the competition assessors.

DESIGN

The exhibition is designed by Nigel Coates of Branson Coates Architecture, London, and aims to present architecture to the public in an accessible, informative and exciting way. Recent designs by Branson Coates include the Arca di Noe restaurant and the Art Silo, both in Japan, and the practice has been selected to design a new wing for the Geffrye Museum, London. In 1995 Nigel Coates was appointed Professor of Architectural Design at the Royal College of Art.

SPONSORSHIP

As local authority for the City of London, the Corporation's substantial sponsorship of this exhibition is totally in keeping with its deep involvement with London as the world's financial and business capital: and also with the Corporation's centuries old trusteeship of probably the world's most famous inhabited Bridge, Peter de Colechurch's London Bridge, and its modern successor and neighbours at Tower, Southwark and Blackfriars.

As a world leader in utility and community services, the Générale des Eaux Group is also pleased to support this exhibition. Their businesses seek to maintain close ties with the local communities where they operate and to bridge the divide between private and public service, thus making this support all the more pertinent.

CATALOGUE

A fully illustrated catalogue will accompany the exhibition, and is available in paperback from the Royal Academy, and is published in hardback by Prestel.

EDUCATION

An extensive programme of educational activities has been designed to make the exhibition accessible to the widest possible audience. In particular, a programme called **Building Bridges** will bring children from schools in the London Boroughs of Brent, Hackney, Lambeth, Lewisham and Southwark to explore the role of bridges in our cities. Separate release available.

DATES

Press Day:	**Tuesday 24 September 1996, 10am - 4pm**
Private View:	Wednesday 25 September 1996, 10am - 8.30pm
Open to Public:	Thursday 26 September - 18 December 1996

HOURS OF OPENING

10am - 6pm daily, including Sundays (last admission 5.30pm).

ADMISSION

£5 full charge; £3.50 concessions; £2.50 12-18 years and £1 8-11 years.

For further press information please contact Katharine Jones, Press and Promotions Officer, or Linda Weston, Deputy Press Officer on tel: 0171-494 5610/5615 or fax on: 0171-439 4998.

updated 23.9.96

ROYAL ACADEMY OF ARTS PRESS RELEASE

PICCADILLY LONDON W1V 0DS TEL: 0171 494 5615 FAX: 0171 439 4998
REGISTERED CHARITY NUMBER 212798

THAMES WATER HABITABLE BRIDGE COMPETITION

A limited competition, supported by Thames Water Plc, to design an inhabited bridge for the River Thames, was launched by the Royal Academy of Arts in April.

Seven international architectural practices were invited to submit designs, and their proposals offer radical re-interpretations of the idea of an inhabited bridge. "The solutions prepared by the architects show that a new inhabited bridge in London is a practical reality", said Peter Murray, organiser of the competition. "We are now expecting developers to come forward who are willing to take this exciting project onto its next stage. We have already received a number of positive enquiries."

Two designs, by Antoine Grumbach & Associates and Zaha M. Hadid, were selected as joint winners by a panel of assessors, but visitors to the exhibition will be able to vote for their favourite design from the following range of dramatic and individual schemes:

Antoine Grumbach & Associates (France) "The Garden Bridge" - an arcade of shops and café spaces placed between two towers on the north bank and a glass palace on the south. Antoine Grumbach won the 1992 French National award for excellence in urban design, and is currently working on an important redevelopment of the metro and rail station at Tolbiac in Paris.

Architectburo Libeskind (Germany) Libeskind deconstructs the concept of a bridge to create a network of paths across the river. Within the net are stopping points, pavilions, cafés and viewing points. Daniel Libeskind recently won the design competition for the Boilerhouse extension at the Victoria and Albert Museum, London.

Branson Coates (UK) Branson Coates's anthropomorphic design houses a 24 hour leisure and entertainment centre with two hotel towers close to the South Bank. Branson Coates has been selected to design a new wing for the Geffrye Museum, London. Nigel Coates is Professor of Architectural Design at the Royal College of Art.

Future Systems (UK) "The People's Bridge" - a pedestrian bridge with space for shops and restaurants, this proposal uses shipbuilding technology to create a curvilinear form. Future Systems recently completed a floating bridge in London's Docklands and are working on the New Media Centre at Lords Cricket Ground.

Ian Ritchie Architects (UK) The top level of Ritchie's proposal is a park; beneath, there are bowling alleys and cinemas as well as viewing platforms. Ian Ritchie Architects are currently designing an extension to the Museum of London, and the International Rowing Centre in London's Docklands.

Krier Kohl (Germany) The most traditional of the entries, this comprises a gothic style façade spanning the river between two towers of residential accommodation. Krier Kohl's speciality is urban planning, and they recently planned the redevelopment of the inner-city station area around The Hague, Netherlands.

Zaha M. Hadid (UK) The design cantilevers accommodation from the banks, while allowing space for views through the central section. Zaha Hadid won the Cardiff Bay Opera House design competition last year, and is currently working on an exhibition installation for the Vienna Kunsthalle.

The competition has been supported by the Secretary of State for the Environment, the Rt. Hon. John Gummer MP, who is eager to see a new bridge constructed across the River Thames. On launching the competition in April, Mr Gummer commented:

> "It is well known that I wish to see the River Thames playing an even greater role in the life of this city. This initiative by the Royal Academy of Arts to hold a design competition for a new bridge will give a real boost to the development of the river. This proposal for a new London bridge will catch the imagination of London."

SPONSORSHIP

Sir Robert Clarke, Thames Water's Chairman, said: "Since 1989 Thames Water has invested £350 million at sewage treatment works to improve the quality of waste water returned to rivers. As a result of this programme the River Thames is the cleanest it has been in living memory. Supporting the Royal Academy of Arts bridge competition seemed to be an ideal way of celebrating this achievement."

The assessors were Sir Philip Dowson PRA (Chairman), Rt. Hon. John Gummer MP, Michael Cassidy (Corporation of London), Sir Robert Clarke (Thames Water), Jean Dethier (Centre Georges Pompidou), Gordon Graham PPRIBA, Peter Murray (exhibition curator) and Janet Street-Porter (broadcaster).

For further information or photographs please contact Katharine Jones, Press and Promotions Officer, or Linda Weston, Deputy Press Officer on 0171-494 5615/5610 or fax on 0171-439 4998.

20.9.96

THE CITY AS A NEW CONCERN FOR CULTURAL EXHIBITIONS.

« INHABITED BRIDGES » AS A PILOT PROJECT

From its inception in 1977, the *Centre Georges Pompidou* in Paris has deliberately presented its visitors (7 millions per year) with a multidisciplinary approach to 20th century art and culture. Hence, architecture has formed the subject of over 140 exhibitions. During this period Jean Dethier has contributed greatly to this on-going, innovative policy which has attracted large audiences. However, since the early 1990s, he has concentrated on exhibitions focused on the city and on urban design. As architecture has gradually established a new « right to be heard » in museums over the past twenty years, it is now important to recognise that the city, as a major facet of modern culture, has been almost entirely neglected as a legitimate theme for exhibitions. It was in order to counteract this cultural amnesia that Jean Dethier pioneered the first major interdisciplinary exhibition in Europe on the theme of « *The Modern City* » which was shown at the *Centre Pompidou* in 1994 and then in Barcelona and Tokyo in 1996. It was with the same innovative intent that in 1990 he began work, in collaboration with Ruth Eaton, on an exhibition dedicated to a building type which frequently represented the spirit of urban vitality : the « *inhabited bridge* » in the past, present and future.

AN EXCEPTIONAL EVENT

The exhibition « *Living Bridges* » which was thus initiated in Paris by the Centre Pompidou and completed by the *Royal Academy of Arts* in London, constitutes an exceptional event on several counts. It explores a fascinating yet little known topic which has never been studied before in any depth and hence can be considered a trailblazing event. The exhibition presents a creative interpretation and a vigorous application of that interdisciplinary approach so central to the Centre Pompidou's mission since, by definition, the inhabited bridge assumes a synergy between three disciplines which are rarely reconciled : engineering, urban design and architecture. The exhibition provides a coherent on-going study of the many different stages and facets of the evolution of this archetype, from its medieval origins to its most contemporary manifestations. In addition, it has very real implications for the issues of today. By proposing to resurrect the inherited wisdom of inhabited bridges it is able to establish constructive proposals for the modernisation of a building type relevant to the requirements of cities and citizens now and in the future.

COMBINING EXHIBITION AND COMPETITION
TO STIMULATE NEW FORM OF CIVIC ACTION

In this respect, the link between the exhibition and the international architecture competition organised in order to encourage the construction of a new inhabited bridge over the River Thames, in the heart of London, represents a stimulating cultural event and allows seven very different entries to be presented to the public, to decision makers and to developers.

An organic relationship between an architectural exhibition and an architecture competition, in order to realise an exemplary project, has twice been explored at the *Centre Pompidou* by Jean Dethier. In association with his exhibition, « *Down to Earth* », he encouraged a development of 72 experimental and ecologically-sound houses built in 1983 by the State in the new town of L'Isle d'Abeau near Lyons. In 1988, in the wake of his exhibition « *Châteaux Bordeaux* », which invested the architecture associated with vineyards with a new value, a private company constructed an innovative winery for *Château Pichon-Longueville* at Pauillac in the Medoc.

These two successful enterprises prove that it is possible for museums and cultural institutions to become directly involved to promote architectural and urban design progressive concepts. This form of civic action, still very rare, reflects the mission of the *Centre Pompidou* as enunciated in its law of foundation passed by the French Parliament. I am delighted that, in the context of the exhibition « Living Bridges » the *Royal Academy* can share this dynamic cultural approach with us, in close cooperation with the British Government represented by the *Secretary of State for the Environment*.

Such projects permit cultural institutions to act less passively in respect of the destiny of our cities, to participate actively in the debate concerning the future of our urban environment. I am delighted that these visionary ambitions could be realised within the context of the first collaboration between two major European cultural institutions : the *Royal Academy of Arts* and the *Centre Georges Pompidou*.

Germain Viatte
Director
Musée national d'art moderne - Cci
CENTRE GEORGES POMPIDOU, PARIS

(extracts of the foreword to the catalogue « Living Bridges »)

GENERALE DES EAUX
G R O U P

PRESS RELEASE

Générale des Eaux Co-Sponsors Living Bridges Exhibition

Water under the bridge: French water and service company floods London's Royal Academy

Générale des Eaux, after more than a century as France's leading water company, has developed unrivalled expertise in the engineering and operation of water supply and treatment services. As part of its sponsorship deal, the multi-utility and services group has designed and constructed a series of tanks, which will transform the Main Galleries of the Royal Academy when they are filled with flowing water.

Living Bridges is being jointly sponsored by Générale des Eaux and the Corporation of London. The idea behind the exhibition, which opens to the general public on 26th September, and runs until 18th December is to show the important role that rivers have played in cities ancient and modern. In this context the river also represents the passage of time itself.

Générale des Eaux was established in 1856 to provide French local authorities with water supply and treatment services. After more than a century as France's leading water company, the group has developed unrivalled expertise in the engineering and operation of water services world-wide. Like living bridges through the ages, it has developed in size, strength, diversity and expertise. Today, 2,500 operating companies employ 215,000 people and turnover £18bn annually. And group activities span the range from civil engineering and construction, waste and energy management, transportation, telecommunications and healthcare.

The Group has a nine-year track record of investment in the UK, and now employs some 25,000 people generating a turnover of nearly £1.5 billion each year. Générale des Eaux' activities in the UK span water supply, telecommunications, waste, energy and facilities management, combined heat and power, healthcare, construction, transportation and parking. It's investment of some £1,000 million in the UK since 1987 demonstrates its long-term commitment, and reflects its intention for the UK to be a central part of Générale des Eaux' overall development.

But despite its size, Générale des Eaux remains essentially a group of small and medium-sized companies which have grown from, and retain very close ties with, the areas where they operate. So, in common with the great living bridges of the past and present, Générale des Eaux has both successfully withstood the test of time and adapted to change.

THE CORPORATE INTERESTS OF THE GENERALE DES EAUX GROUP IN THE UK ARE REPRESENTED BY GENERAL UTILITIES PLC

37-41 OLD QUEEN STREET LONDON SW1H 9JA TELEPHONE +44 (0)171 393 2700 FAX +44 (0)171 393 2748

GENERAL UTILITIES PLC REGISTERED IN ENGLAND No 2127203

A BRIDGE OVER CLEANER WATER

A competition to design a new bridge across the River Thames in London celebrates the river's status as the cleanest metropolitan river in the world, says Thames Water.

Thames Water's support of the Royal Academy of Arts competition commemorates the success of the company's investment programme at sewage treatment works.

Sir Robert Clarke, Thames Water's Chairman, said: "Since 1989 Thames Water has invested £550 million at sewage treatment works to improve the quality of waste water returned to rivers. As a result of this programme the River Thames is the cleanest it has been in living memory.

"Supporting the Royal Academy of Arts bridge competition seemed to be an ideal way of celebrating this achievement".

Salmon are now returning to the Thames in increasing numbers after an absence of 160 years. In total the river is home to over 100 species of fish. In April 1996 an otter was sighted in the River Thames near Reading. Last year 11 groups who use and monitor the River Thames also presented Thames Water with a certificate endorsing the company's contribution to cleaning up the river.

Sir Robert Clarke added: "Our investment over the past seven years has paid real dividends, but our task is not yet finished. By the year 2000 we plan to spend well over £300 million at sewage treatment works to raise standards even higher."

Each day Thames Water's 365 sewage treatment works treat 3,959 million litres of waste water from the company's 11 million customers.

26 September 1996   ends

For further information please contact Thames Water Press Office: Nick Tennant, 01734 593396, Chris Foreman, 01734 593364, Samantha Poë, 01734 399258 or Jane Frapwell, 01734 399271.

Thames Water Utilities Ltd Nugent House Vastern Road Reading RGI 8DB
Direct Telephone lines: (01734) 593396 / 399258 Fax: (01734) 599295

6477A

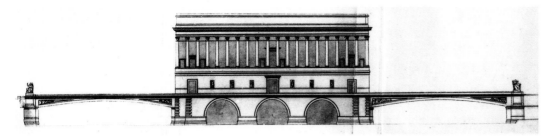

Fig. 125,
L. P. Baltard, *Project for a bridge with law courts and a prison to be built on the Pont au Change: elevation*, c. 1828, lithograph. Bibliothèque Municipale de Lyon, Coste 666

conceived on a grand scale, with certain of its features reminiscent of du Cerceau's design two centuries earlier for the Pont Neuf site in Paris (see pp. 60, 61, fig. 72). Baltard proposed the creation of two new squares, one for the court house and the other for the prison, supported on an enormous plinth-bridge to be built over three tunnels aligned with the flow of the river. While the alignment of the original Pont du Change would provide access to the court house, a second, new bridge would provide access to the prison. Lateral arcades on the ground floor were to provide commercial accommodation, and a colonnade on the first floor of the court house was to act as a promenade for the magistrates (figs. 125-127). Other drawings, now housed in the Archives Municipales, Lyon, show that Baltard also considered the site appropriate for a building incorporating shops and a theatre (figs. 123, 124).

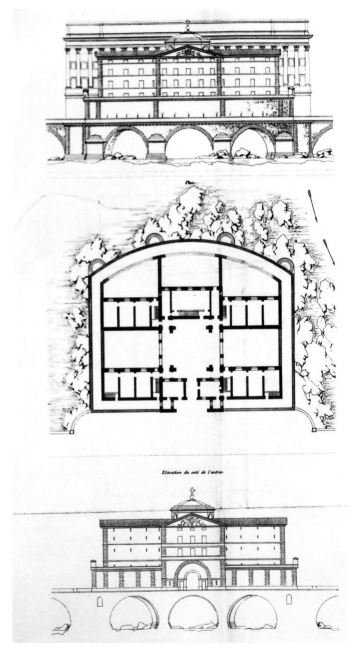

Fig. 126,
L. P. Baltard, *Project for a bridge with law courts and a prison: back elevation, plan, entrance elevation of the prison*, c. 1828, lithograph. Archives Municipales, Lyon

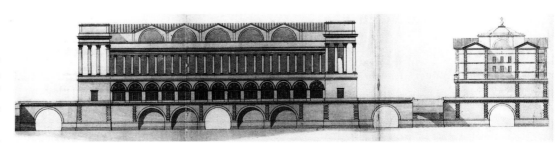

Fig. 127,
L. P. Baltard, *Project for a bridge with law courts and prison: cross-section*, c. 1828, lithograph. Archives Municipales, Lyon

Fig. 128, Thomas Mosley, *Proposal for a bridge with a European art gallery to be built over Waterloo Bridge, London*, published in the *Pictorial Times*, 1843. The British Library

MOSLEY'S PROPOSAL FOR WATERLOO BRIDGE, LONDON

In 1843, the *Pictorial Times* printed an explanatory text accompanied by illustrations (fig. 128) for a proposal by Thomas Mosley for a 'European Universal Gallery'. It was to be built over the recently constructed Waterloo Bridge. The text reads as follows:

'Our readers have now before them engraved copies of some drawings illustrating a proposed change, of novel and somewhat startling character, in the finest of our metropolitan bridges. The designs have been submitted to Prince Albert and other persons of taste and influence, and there is every probability of the plan being realised. The first sketch represents the elevation of a structure proposed and designed by Mr. Thomas Mosley, civil engineer, Bristol, to be erected over the whole length and breadth of Waterloo Bridge, constituting a room or gallery, divided into suitable compartments,

1240 feet long, and 42 feet wide, with an uninterrupted promenade in the middle of the room, 12 to 15 feet wide, the whole length of the building. It is also proposed to construct a conservatory over the room, extending the length of the three centre arches, about 400 feet long, 42 feet wide, and from 12 to 15 feet high, with plate glass fronts, and a promenading room at each end 18 feet by 42. The part of the room over the three side or end arches is proposed to be lighted by skylights, consequently the apparent windows in that portion of the building are blank, whilst those under the conservatory will be glazed with plate glass. The fabric will be supported either entirely by cast iron pillars and arches, or a combination of stone and iron, namely, those on the outside may be granite, and the interior row near the edge of the pavement or footway be of hollow cast iron, so that the building is intended to rest on four rows of pillars. The

whole structure above the arches to be composed of wrought iron framing, with proper diagonal stays and ties. The beams and joists to support the floor and roof, to be also drawn wrought iron, and the flooring is proposed to be a species of tessellated pavement. The exterior and interior of the building to be covered or coated with fire-proof cement, so as to produce the appearance and durability of stone, with perfect safety from the accident of destruction by fire. For security against the effect of wind, it is proposed to employ numerous strong wrought iron bolts to pass through the hollow iron pillars and into, and in some places through, the masonry of the bridge, firmly keyed to cast iron plates in the foundation and at the under part of the massive arches. The room or gallery is proposed to be appropriated to the exhibition and sale of works of art, science, and literature, from all parts of the world, and be denominated by the European Universal

Gallery. It is understood that Mr. Mosley has arranged with an eminent London architect for another design, varying the style of architecture so as to harmonise in a still higher degree with the present elegant structure.

The foregoing description is in the words of the projector, and we preferred giving it in that shape to enlarging in any way upon the plan he proposes. The undertaking is an extensive one; but as the bridge has hitherto, in a monetary point of view, been a failure, it is more than probable that the projected changes will be made, since the rent of the proposed arcade would be a source of permanent revenue'.

The use of cast iron and glass for the construction of various forms of inhabited bridge was explored extensively throughout the nineteenth century. While often not as ambitious as Gustave Eiffel's proposed superstructure over the Pont d'Iéna, Paris (see p. 90), the materials could be used for a bridge-market complete with stalls proposed across the River Limmat in Zurich in 1823 (fig. 129), or for the provision of kiosks upon a suspension bridge, as suggested by Mr Alcock MP in *The Builder*, 1853, for a 'Bazaar Bridge' (fig. 131). Alcock specifically quoted the example of Florence's Ponte Vecchio (see pp. 62-65) as a prototype for his 'bazaar bridge', and cast iron and glass were also proposed by Martelli and Corazzi when they came to address the modernization of that bridge after 1850 (fig. 130).

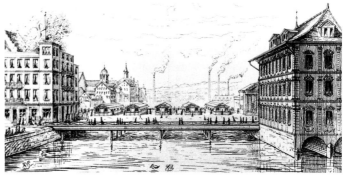

Fig. 129, Rathausbrücke, Zürich: *Project for a market bridge*, 1823-24, engraving. Baugeschichtliches Archiv der Stadt Zürich. The project proposed to place five markethalls on a cast iron bridge over the River Limmat.

Fig. 130, Giuseppe Martelli, *Proposed reconstruction of the Ponte Vecchio, Florence*, c. 1850, colour lithograph

Fig. 131, *Proposal for a bazaar-bridge*, published in *The Builder*, 1853. British Architectural Library, RIBA, London

GALMAN'S PROPOSAL FOR AMSTERDAM

In 1848, the Dutch hydraulic engineer J. Galman made a proposal for a bridge across the River Ij in Amsterdam (fig. 132). The arguments which he gave in support of his proposal echo those given earlier by William Bridges for his bridge over the Avon Gorge in 1793 (see p. 82, figs. 120, 121) Mullgardt and Hood and in San Francisco and New York respectively in the early twentieth century (see pp. 94, 95, figs. 137-140). Like Bridges, Mullgardt and Hood, Galman stressed the financial benefits of his bridge whose apartments, shops and integral warehouses would generate rental income. In Galman's case he planned to build the bridge at his own cost and then donate it to the municipality in exchange for a percentage of the bridge's rental income and concessions to build new districts in the north of the city.

Situated in the harbour of the River Ij, his concept was technically advanced. He proposed creating two 180-metre-long sand buttresses, which would have the additional advantage of counteracting subsidence within the harbour. Upon these he placed the warehouses, which in turn would support sloping ramps leading up to a central iron bridge some 163 metres in length. The bridge itself was not intended to support buildings, although it did have two pillars at its central point designed in the style of fortified town gates. In a reworking of his project in 1857, Galman suggested adding a double row of shops and lodgings in vernacular style on the sloping ramps.

Galman's project failed to win official approval despite drawing its inspiration from a traditional Dutch building type dating back to the Middle Ages: that of the sluice-bridge. In the medieval city where space within the city walls was limited, such bridges had served as temporary market places and warehouses, with domestic accommodation beneath. In the mid-nineteenth century, however, such constraints on space were no longer relevant and road traffic was insufficient to justify the construction of so costly a new bridge.

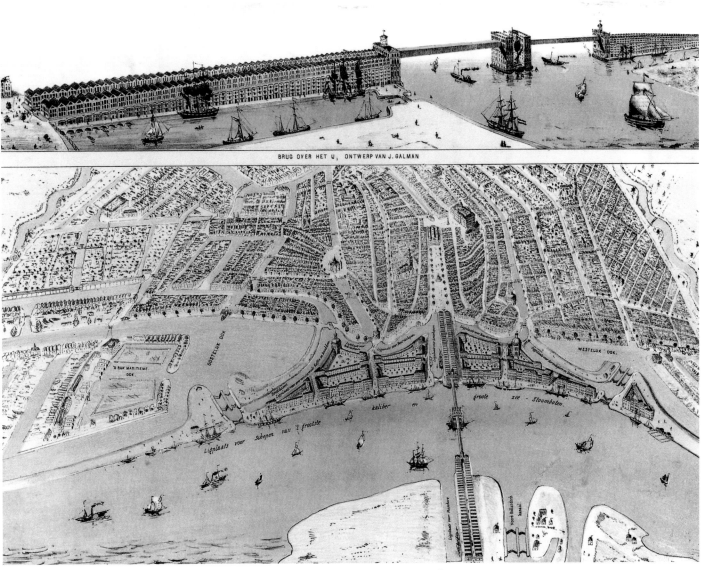

Fig. 132, J Galman, *Proposal for a bridge over the Ij, Amsterdam*, 1857 engraving. Gemeentearchiv Amsterdam

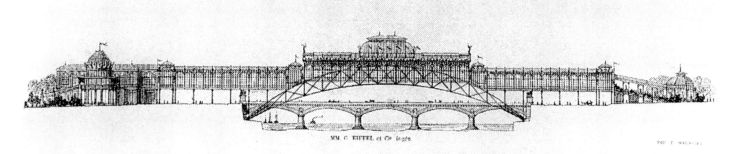

Fig. 133, Gustave Eiffel, *Proposal for a bridge to be built over the Pont d'Iéna*, Paris, 1878, engraving. Bibliothèque des Arts Décoratifs, Paris, Collection Maciet

EIFFEL'S PROPOSAL FOR THE PONT D' IENA, PARIS

Gustave Eiffel understandably lacked modesty when he described his competition entry for a bridge above the Pont d'Iéna for the Paris Exposition Universelle of 1878 as 'original, audacious and of extraordinary technical interest' (fig. 133). Unfortunately, like many other inhabited bridge projects, it was rejected by the authorities, the Prefect of the Seine complaining that its structure would obscure the view along the river towards the newly erected Palais du Trocadero at Chaillot.

Eiffel's striking bridge differs from most other projects reviewed in this catalogue, not least because it was to have been built (like Mosley's proposal for Waterloo Bridge, see pp. 86, 87) above the existing Pont d'Iéna. The structure was light and graceful in appearance for it consisted of a single 130-metre long cast iron arch surmounted by a glass-clad arcade. A staircase and entrance pavilion provided access at each end and the central section contained a large hall measuring 70 by 24 metres. Rather than carrying shops or housing, its function was to provide a space for exhibitions and fêtes and to connect the two main sites of the Exposition Universelle, the Champ de Mars and Chaillot. Unfortunately, little documentation about the project survives but it is reputed to have won the enthusiastic acclaim of Jean-Baptiste Krantz, Commissioner-General of the Exposition Universelle and of the public. One can but speculate that, had the bridge been erected, Eiffel would have been correct in his assessment that 'a work like this can make me the equal of the most important builders in Europe'. Instead, such an accolade would have to remain unclaimed until Eiffel had erected his tower at the succeeding Exposition Universelle, mounted in 1889.

The ambitious nature of Eiffel's design was mirrored in other European projects conceived during the closing decades of the nineteenth century. While Friedrich Keck proposed linking the two banks of the River Rhine in Basel with a glass and iron assembly hall and market (fig. 134a), Tellier produced a radical solution in 1891 to ease transport circulation in Paris (fig. 134b). A metropolitan railway was to be constructed above the River Seine, with its central station being located above the Pont de la Concorde.

Fig. 134 (a), Friedrich Keck, *Project for a double bridge supporting an assembly hall and shops*, c. 1899, lithograph. Staatsarchiv Basel-Stadt

Fig. 134 (b), C Tellier, *Project for a station to be placed over the Pont de la Concorde, Paris, for a proposed metropolitan railway*, 1891, engraving. Bibliothèque des Arts Décoratifs, Paris, Collection Maciet

The Twentieth Century

Fig. 135, William Walcot after Edwin Lutyens, *Proposal for an art gallery over the River Liffey, Dublin*, 1913, pen and ink. British Architectural Library, RIBA, London

LUTYENS' PROPOSAL FOR DUBLIN

In 1913, Sir Edwin Lutyens designed an art gallery on a bridge in Dublin for the art collector Sir Hugh Lane, for whom he had already designed the Johannesburg Art Gallery in 1910.

Lane wished to open a Gallery of Modern Art in Dublin. He had reached an agreement with the City Corporation in 1906 whereby he would lend the main part of his collection for temporary display in the Municipal Gallery on the understanding that it would be moved to a new permanent building on a different site within a few years. Frustrated by the lack of progress in creating the new gallery, and facing opposition to the various locations he had already considered

in Dublin, Lane suggested in 1913 replacing an existing metal bridge over the River Liffey with a stone-clad bridge supporting a new gallery to be designed by Lutyens (fig. 135). The style was in the classical beaux-arts tradition which had been applied to the triumphal bridge projects from Ennemond-Alexandre Petitot to Otto Wagner (see pp. 76-81, figs. 103-109) and was also to be used by Collcutt in his scheme to rebuild Charing Cross Bridge, London (fig. 136). In plan, Lutyens proposed an H-shaped construction with two galleries adjacent to the quays and a smaller gallery surmounted by an open, colonnaded footbridge across the river (the function of passage being subordinate to that of the gallery). Unfortunately the scheme was

Fig. 136, Thomas Collcutt, *Project for a street bridge to replace Charing Cross Bridge, London*, 1906, engraving. British Achitectural Library, RIBA, London

not realized despite its approval in 1913 by the Dublin Corporation, which agreed to cover half the costs. Rising opposition to the scheme created difficulties in raising the balance of the funding from private subscription. Embittered by the arguments put forward against his project, including an announcement by the Dublin Corporation that they had not approved his choice of architect and would have preferred a competition among Irish architects for the same site, Lane withdrew his pictures and lent them instead to the National Gallery in London. Lane's difficulties with his collection did not end there. Bequeathing the works to London in his will, he disagreed with the conditions imposed for accepting them. An unwitnessed codicil to his

will in 1915, again leaving the paintings to Ireland, was considered legally invalid on his death when the Lusitania went down later that year.

Dublin was not to have a bridge-gallery at the beginning of the twentieth century, but the end of the century has seen the revival of the idea in Michael Graves' project for Fargo and Moorhead in the U.S.A. (see pp. 104-105, figs. 157-159) and William Alsop's for a new building for the Institute of Contemporary Arts across the River Thames in London (see pp. 128-130, figs. 215-218).

MULLGARDT'S PROPOSAL FOR SAN FRANCISCO

In 1924, shortly before Raymond Hood made his proposal for New York (p. 95, figs. 139-140), the San Francisco architect Louis Christian Mullgardt designed a visionary 'skyscraper bridge' linking San Francisco and Oakland (figs. 137-138). This was radically different from most previous inhabited bridges – apart from William Bridges' design for a bridge over the Avon Gorge, (see p. 82, figs. 120, 121) in that all the residential and commercial space was placed not above the roadway of the bridge itself but in its pylons. At the time, around a decade before the Bay Bridge and Golden Gate Bridge crossed the bay, even the idea of spanning such a distance seemed fanciful and Mullgardt's scheme was given serious consideration neither by investors nor by the planning authorities. Moreover, the War Department blocked his scheme on the grounds that the fleet would be trapped at its anchorage were the bridge to collapse.

Mullgardt's bridge was to have been composed of a steel structure with ten steel piers each 285 metres apart serving as buildings with rentable interior spaces. The three central piers would be the highest, rising about 86 metres above the water level to enable the passage of large ships. Platforms for passenger boats would extend out from the concrete foundations at right angles, allowing passenger boats to dock, and seaplanes and Zeppelins to land. The causeway above the bridge was initially to have one level, with an option to add double or treble decking if required. The pylons would provide prime space: easily accessible, with good views, lighting and fresh air, they could be used for housing, offices, hotels, auditoriums, factories and garages while the arches could serve as hangars. Lifts would transport passengers directly from the terminals below into the pier buildings or on to the causeway for connection with other means of transport.

Although his project had certain technical failings it was not unrealistic, and Mullgardt argued in *The Architect and Engineer* of March 1927 that it could also have been profitable: 'This form of bridge construction is serviceable for every desirable purpose: its realm of utilization is unlimited and most economical. To ultimately convert the bridge supports into housing, admirably suited to every conceivable purpose, is not only feasible, but logical. It can be done at a minimum cost. The ground areas upon which they stand do not require to be purchased: the foundations are built: they rest upon earth which is tax free. The steel frame is practically complete. The actual cost of such buildings is determined by wall, windows, doors, partitions, elevators, plumbing, electric wiring and whatever else shall be required to meet the lessee's needs'.

Fig. 138, Louis Christian Mullgardt, *A multiple bridge for San Francisco Bay between San Francisco and Oakland*, 1924, published in *The Architect and the Engineer*, vol 88, March 1927

Fig. 137, Louis Christian Mullgardt, *A multiple bridge for San Francisco Bay between San Francisco and Oakland: cross-section*, 1924, published in *The Architect and the Engineer*, vol 88, March 1927

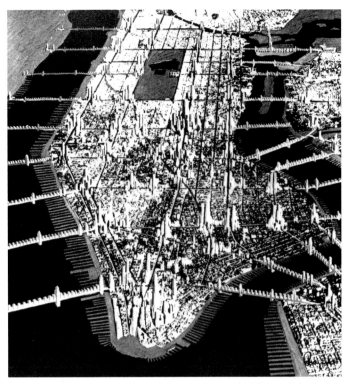

Fig. 139, Hugh Ferriss after Raymond Hood, *Apartments on a bridge*, published in Hugh Ferriss, *The Metropolis of Tomorrow*, 1929. New York Public Library

Fig. 140, Raymond Hood, *Manhattan 1950*, 1925, reworked for an exhibition at the Architectural League, New York City, 1930. New York, Public Library

HOOD'S PROPOSAL FOR NEW YORK

It would appear that the inhabited bridges for New York proposed by Raymond Hood and illustrated by Hugh Ferriss in the 1920s remain the most ambitious and megalomaniacal ever produced (figs. 139-140). Not only was the scale of the blocks of housing and other amenities upon the bridge impressive but Hood's vision, which he reworked several times, went even further: he imagined up to a hundred such bridges linking Manhattan to its boroughs and New Jersey. Its extensive urban dimension recalls, on a vastly expanded scale, pre-eighteenth century Paris, with its numerous bridges around the Île de la Cité and the Île Saint-Louis (see pp. 52-61).

Before his early death in 1934, Hood was responsible for five of New York's most important skyscrapers: the Chicago Tribune building (with John Mead Howells), the American Radiator Co. building, the Daily News building, the McGraw-Hill building and the RCA building (with the Associated Architects of the Rockefeller Center). Hood's first design for a bridge city was published in *The New York Times* of 22 February 1925, in an article entitled 'Bridge Homes: A New Vision of the City', and depicted in a striking perspective by Hugh Ferriss, a talented architectural delineator and visionary. Some 50,000 people would be housed in the massive skyscraper piers and other lower skyscrapers under the arcs of the suspension cables along the causeway of Hood's bridge, which would also support shops, theatres, esplanades and roof gardens. Ferriss included drawings of the 'Apartments on Bridges' concept in his 1925 exhibition *Drawings of the Future City* and his 1929 book *The Metropolis of Tomorrow*. Hood reworked his scheme in 1926 on a grander scale, proposing twenty, fifty or even one hundred foot-long spans in an article in *Liberty* magazine and again in 1930 for the annual exhibition of the Architectural League of New York under the title '*Manhattan 1950*'. Although these schemes were not realized, they made some economic sense. Hood claimed that, with prime Manhattan property valued at $3,000 or more per foot of frontage, the value of the real estate on each bridge could be estimated at around $60 million. This recalls the financial arguments put forward by Louis Christian Mullgardt for San Francisco around the same period (see p. 94): both architects had developed the notion of air rights but their proposals were to fall on deaf ears.

MORGAN'S PROPOSAL FOR CHICAGO

In 1928, readers of the *Chicago Sunday Tribune* learnt of a proposal by the architect and artist Charles Morgan and the architectural practice of D.H. Burnham & Co. for an inhabited bridge for Chicago. The accompanying article by Al Chase provided the following information about this project which, like those of Mullgardt for San Francisco and Hood for New York (see pp. 94-95), was not to be realized:

'Rainbow bridge, a huge arched army of skyscrapers in a single row, bending in a sweeping band of color from Lake Shore drive to Randolph Street, is the suggested method of linking north and south sides near the mouth of the river, made by Charles Morgan, Chicago artist and architect. The location and application of the project to the city's development is being made by D.H. Burnham & Co. The name of the great proposed structice [sic], Rainbow Bridge, is given because of the sweeping bands of color suggested by Mr. Morgan which would result in a rainbow effect when seen from the lake or viewed from Michigan Avenue.

The startling novelty in Mr. Morgan's project is the utilization of the great piers as office building skyscrapers of twenty-five or more storeys. Still more novel would be the entrance into these unique structures – from the top. The tenants would whiz over the bridge to the pier which contained their respective building, park their cars in the proposed garage in the upper floors, and drop down in an elevator to their offices'.

A width of approximately 150 feet is suggested for the bridge by the architects. The grade would be about 8 per cent.

Fig. 141,
Charles Morgan with D. H Burnham & Co, *Chicago: proposal for a rainbow bridge*, published in the *Chicago Sunday Tribune*, 1928.
The British Library

MELNIKOV'S PROPOSAL FOR PARIS
FUMET & NOIRAY'S PROPOSAL FOR LYON

In the 1920s, while the American public was being presented with schemes for residential bridges on a grand scale in San Francisco, New York and Chicago (see pp. 94-96), two proposals were put forward in France for garage-bridges.

In February 1930, Fumet & Noiray, hydraulic engineers, put forward a proposal, accompanied by a drawing, for a concrete garage-bridge over the River Rhône opposite the Place de la Charité in Lyon fig. 142). It consisted of a bridge about 278 metres long and 38.5 wide; the north side would contain a roadway for vehicles and two pedestrian pavements, and the south side would support an eight-storey garage. There would be thirty-six shops on the ground floor and room for 216 cars and four offices on each of the seven upper floors. The building, argued the engineers, would contribute greatly to the alleviation of traffic congestion and would be designed in a 'grand style moderne'. The bridge would be offered to the city in return for a 99-year lease on the garage. The City's Department of Public Works turned down the proposal on the grounds that, although it appeared technically and financially feasible, it would block the view along the river banks. In the hope of reversing this refusal, Fumet & Noiray suggested adding a covered swimming pool with a semi-circular 'beach' next to their garage-bridge, but the city authorities remained intransigent.

Five years earlier, and in a much more radical architectural style, the Russian avant-garde architect, Konstantin Stepanovich Melnikov, responded to an invitation from the City of Paris to design a garage for 1,000 automobiles (fig. 143). His unusual suggestion that it should be built on a bridge over the River Seine received, surprisingly, a favourable response. He developed his proposal further, presenting two variants, a minimal and a maximal version. In the former, the river façades of the building were arranged into decorative squares, some of which were left open in order to stress its function through the display of the automobiles and the ramps within the structure. In the latter, he removed the closed façades and proposed a monumental sculpture composed of a series of exposed ramps supported by four piers embedded in the river. The huge caryatids which can be seen in the drawings of this variant served no functional purpose and were merely added by Melnikov as a 'private joke', providing the illusion that they offered additional support to a building which – on account of its structural audacity – appeared precarious. However, neither version inspired the Parisian city fathers to pursue the concept any further.

Fig. 142, Fumet & Noiray, *Lyon: proposal for a garage-bridge over the River Rhône opposite the Place de la Charité*, 1930. Archives Municipales de Lyon

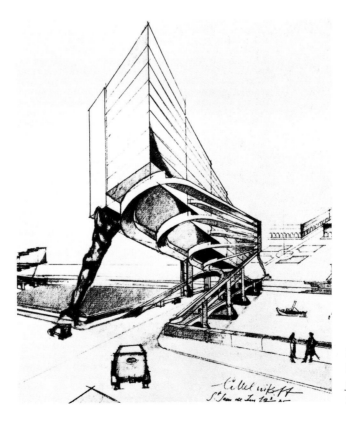

Fig. 143,
K. S. Melnikov, *Paris: proposal for a garage-bridge over the River Seine (revised scheme)*, 1925

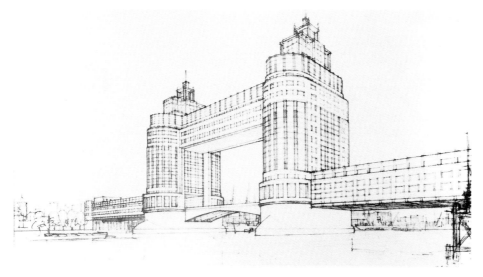

Fig. 144,
W. F. C. Holden, *The Crystal Tower*
Bridge: suggested reconstruction-perspectival view,
1943, pen and ink. © Guildhall Library,
Corporation of London

HOLDEN'S PROPOSAL FOR TOWER BRIDGE, LONDON

London's Tower Bridge is not, strictly speaking, an inhabited bridge as it did not accommodate functions other than those related to the passage of traffic (fig. 145). Its two huge neo-Gothic towers recall the defensive role of fortified bridges such as that at Cahors in France (see p. 42, fig. 43). In reality, however, their purpose was to house the complex machinery initially required to work the opening and closing of the bascules to allow the passage of large ocean-going vessels.

Nonetheless, it is an important structure since its striking architectural aspect enabled it to act as a dynamic link between the two banks of the Thames, to provide a focal point for the Pool of London and to play a symbolic role as an image of London. The problem of erecting a new bridge near the Tower of London was first considered in the 1870s, when over a million people lived to the east of London Bridge,

the most easterly of London's bridges, but were obliged to cross the river by rowing-boat or subway. Numerous projects were elaborated for a new bridge which would allow for the passage of both pedestrians and vehicular traffic and enable large ocean-going vessels to continue their journey upstream as far as London Bridge. The bridge that we know today was designed by the City Architect, Sir Horace Jones and completed by the engineer Sir John Wolfe Barry in 1894.

Tower Bridge received bomb damage during World War II. Rather than repairing it, W. F. C. Holden put forward an unsolicited proposal to the Bridge House Fund in 1943 for a 'Crystal Tower Bridge' (figs. 144, 146). His idea was to replace the bridge with a huge glass superstructure which could house almost 24,000 square metres of office space. The project, however, failed to inspire serious interest and Tower Bridge was merely restored to its previous state.

Fig. 145, William Wylie, *The Opening of Tower Bridge*, 1894, oil on canvas. © Guildhall Library, Corporation of London

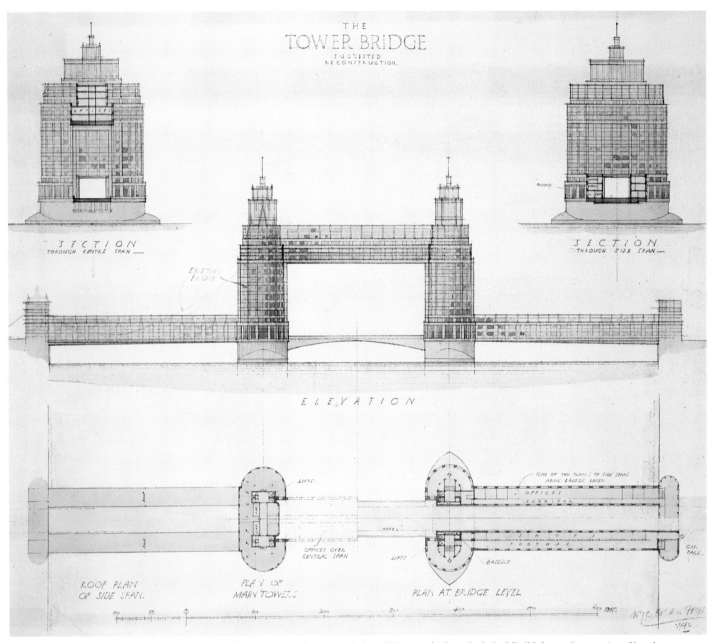

Fig. 146, W. F. C. Holden, *The Tower Bridge: suggested reconstruction-elevations and plan*, 1943, pen and ink wash. © Guildhall Library, Corporation of London

JELLICOE AND COLERIDGE'S PROPOSAL FOR VAUXHAL BRIDGE, LONDON

The Glass Age Development Committee – set up by Pilkington Brothers in order to investigate and promote the use of glass in building – commissioned Jellicoe and Coleridge in 1963 to design a habitable bridge across the River Thames on the site of the existing Vauxhall Bridge (fig. 147).

The proposed Crystal Span comprises a horizontal structure which is supported by two piers in the river and overhangs the banks at either end. It is subdivided into three layers: base, superstructure and roof. At base level run two three-lane carriageways; above this is a service road with parking, approached by ramps from either bank. The superstructure is encased in an air-conditioned glass box providing a total of 290,000 square feet (89,230 square metres) of covered floor space (fig. 149). While this space would have been suitable for residential, office or leisure use, 'the civic nature of the site', wrote the committee, 'must preclude a purely commercial building'. The proposals therefore included a gallery of modern art on the northern pier

(fig. 150), balanced by a luxury hotel on the southern pier. In the centre is a skating-rink, while the base of the superstructure is given over to retail – 'in the form of a modern bazaar' – connected by escalators and a moving foot-way to either bank (fig. 151). The roof is laid out as a series of gardens, with sheltered courtyards and viewing platforms; in the centre is an open-air theatre (fig. 148).

The cost of the whole structure in 1963 was calculated at £7,000,000 and its promoters reckoned that if it were a purely commercial building, it would produce a return of 10 per cent per annum on the capital expenditure.

The length of the bridge is 298.5 metres, the width 38.7 metres and its height 50.7 metres above high water.

Jellicoe and Coleridge, Architects,
Ove Arup and Partners, Engineers 1963

Fig. 147, Jellicoe and Coleridge Architects with Ove Arup, *Crystal Span – a multi-purpose bridge: photomontage*, 1963, Ove Arup Archive

Fig. 148, Jellicoe and Coleridge Architects with Ove Arup, *Crystal Span Bridge: view of open air theatre*, 1963. Ove Arup Archive

Fig. 149, Jellicoe and Coleridge Architects with Ove Arup, *Crystal Span Bridge: section of the northern part, A. The main roadway, B. Service road with parking, C. Shopping arcade, D-H. Three of the seven interlocking halls of the sculpture gallery, I. Viewing platform on the roof*, 1963. Ove Arup Archive

Fig. 150, Jellicoe and Coleridge Architects with Ove Arup, *Crystal Span Bridge: view towards the sculpture gallery and the River Thames*, 1963. Ove Arup Archive

Fig. 151, Jellicoe and Coleridge Architects with Ove Arup, *Crystal Span Bridge: view down the pedestrian walkway*, 1963. Ove Arup Archive

YONA FRIEDMAN'S PROPOSAL FOR PARIS

The 'Paris Spatial' project of 1960 was developed to create air-rights neighbourhoods on 'new' land, without affecting the existing city. Friedman identified railway yards (about 25 per cent of the city area), roads and the river as potential sites.

The megastructure bridges (figs. 152, 153) used a standard space frame, which could accommodate homes, offices or leisure facilities as well as urban motorways. The inhabitants would have been able to select the design of their own homes within the voids created by the structure as long as they followed simple rules about access, natural light and ventilation. The infill was intended to be changeable and expendable; only the structure was to be permanent.

Friedman, one of the best known of the 1960s megastructuralists, was ahead of his time. By the mid-1980s many large-scale projects had been built across railway air rights.

Fig. 153, Yona Friedman, *Paris Spatial Scheme: Paris over the River Seine*, pen and ink, 1960. © Collection Yona Friedman

GUNTHER FEUERSTEIN'S PROPOSAL FOR SALZBURG

Feuerstein's utopian megastructure of 1966 is not strictly a habitable bridge; the fact that it crosses a river is incidental in its march across the city (figs. 154-156). The project forms part of a body of work produced by megastructuralists during the 1960s (see p. 102). The use of living pods, electronic signage, giant moving screens and strong diagonal link elements can also be seen in the work of both the Archigram Group's 'Plug-in City' and the Japanese Metabolists.

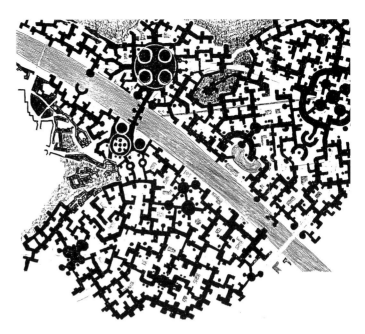

Fig. 154,
Gunther Feuerstein, *Salzburg Megastructure: site plan*, pen and ink, 1966.
© Collection Gunther Feuerstein

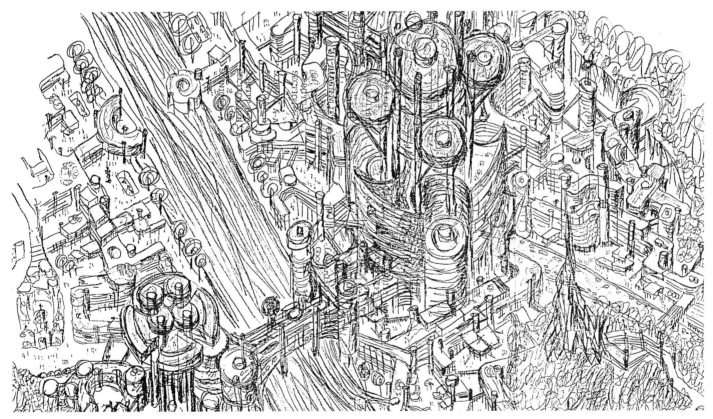

Fig. 155, Gunther Feuerstein, *Salzburg Megastructure: elevation*, pen and ink, 1966. © Collection Gunther Feuerstein

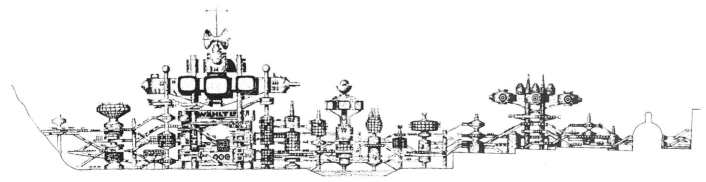

Fig. 156, Gunther Feuerstein, *Salzburg Megastructure: cross-section*, pen and ink, 1966. © Collection Gunther Feuerstein

103

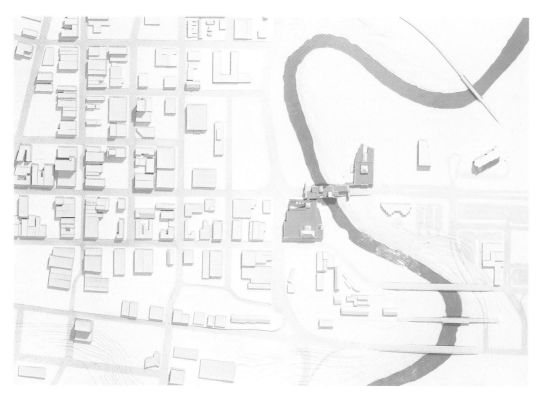

Fig. 157,
Michael Graves,
*Fargo and Moorhead Cultural Center
Bridge, North Dakota and Minnesota:
overhead view of site model*, 1977.
© Collection Michael Graves

MICHAEL GRAVES' PROPOSAL
FOR FARGO AND MOORHEAD

The design for a new cultural centre joining the twin cities of Fargo and Moorhead across the Red River of the North was never built (fig. 157). However, the project, designed in 1977, was important in establishing Michael Graves' position within the Post-Modern pantheon.

An art museum forms the bridge creating the link across the river between the concert hall, and radio and television stations on the one side, and the history museum on the other (fig. 159). The design owes much to Labrouste's beaux-arts proposal for a bridge linking Italy and France, which Graves had seen in Arthur Drexler's exhibition at The Museum of Modern Art, New York, in 1975. According to Graves, the composition attempts 'a vertical unity... by employing the river itself as the basement story, the vehicular access and the first level bridge as piano nobile, and the art museum above the bridge as attic. The horizontal linking members, which are covered aerial walkways connecting the three cultural facilities, are seen as the cornice line of a continuous building' (fig. 158).

The bridge employs a symbolic keystone, which is in fact a window. It also represents a scoop, which collects rainwater from the sky and replenishes the river below through a waterfall which issues from its base. The water is pumped from the river by a windmill, which is part of the history museum and reflects the agrarian base of the local community. Thus the individual elements of the composition are seen as parts of a larger narrative.

Fig. 158, Michael Graves, *Fargo and Moorhead Cultural Center Bridge, North Dakota and Minnesota: elevation*, pencil, 1977. © Collection Michael Graves

Fig. 159, Michael Graves, *Fargo and Moorhead Cultural Center Bridge, North Dakota and Minnesota: south view of site model*, 1977. © Collection Michael Graves

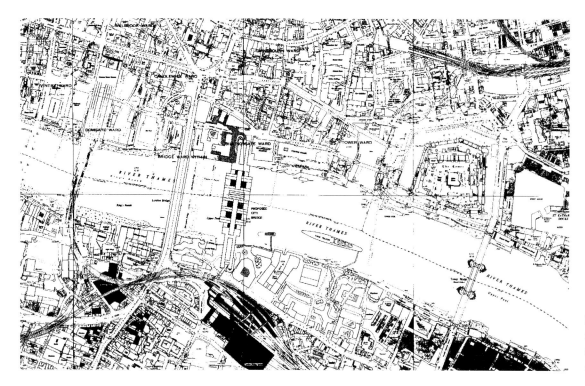

Fig. 160,
R Seifert and Partners,
City Bridge: location plan, 1980.
© Collection J. Seifert

SEIFERT'S PROPOSAL FOR A CITY BRIDGE, LONDON

The City Bridge of 1980 was a serious proposal for a commercially viable new crossing of the River Thames. Designed by John Seifert, it would have provided a direct and amenable pedestrian link from the City to London Bridge Station for the 30,000 people who traditionally used London Bridge every weekday (fig. 160). While the scheme gained financial backers, its visual intrusiveness would have militated against it receiving planning consent (fig. 161).

Spanning the Thames from Billingsgate Market on the north bank to Tooley Street on the south, the bridge was to contain a variety of public facilities, shops and offices (figs. 162, 163): 'The fundamental design principle is that the bridge should not purely be seen as a method of public transportation, but as a raised street full of fascinating and various attractions which will enrich the life of London and draw people to the Thames' (Seifert). At the heart of the proposal is a public piazza the size of Leicester Square, with trees, landscaping, an ice-rink, street cafés and entertainment. Two restaurants high above the square are reached by wall-climbing lifts providing spectacular panoramas of the river. The facilities are serviced by electric trolleys on the bridge deck, which also provides an unobstructed route for emergency vehicles. The bridge contains 14,840 square metres of floor space devoted to shopping and entertainment, 15,300 square metres of public space and 100,000 square metres of space for offices and residential accommodation.

Fig. 161, R Seifert and Partners, *City Bridge: elevation*, pencil, October 1980. © Collection J Seifert

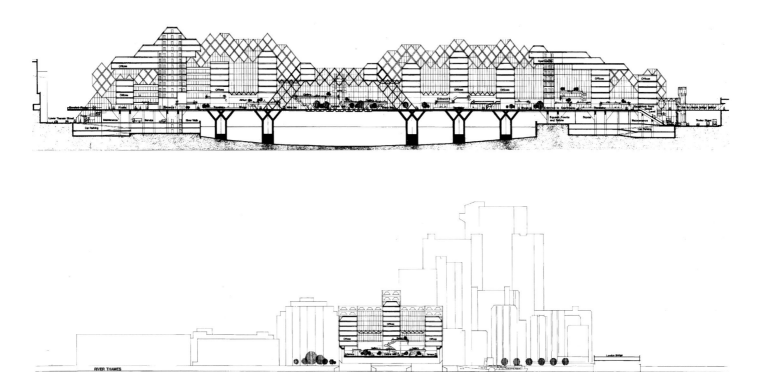

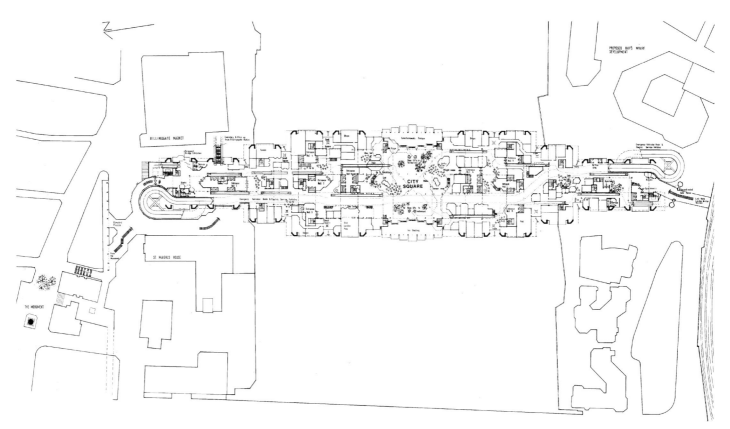

Fig. 162, R Seifert and Partners, *City Bridge: cross-sections*, pencil, 1980. © Collection J Seifert

Fig. 163, R Seifert and Partners, *City Bridge: plan, city square level*, 1980. © Collection J Seifert

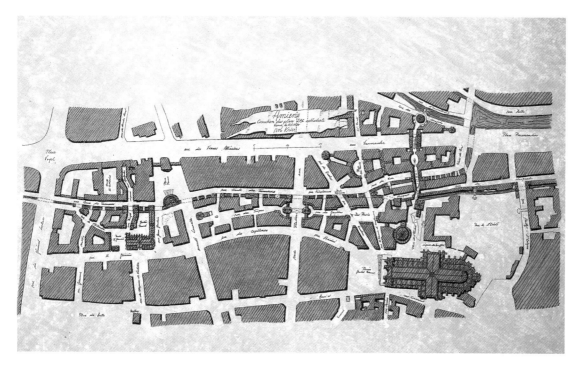

Fig. 164,
Rob Krier, *Amiens
project: site plan-II*, 1984.
© Collection Rob Krier

ROB KRIER'S PROPOSAL FOR AMIENS

Rob Krier is an architect who believes that architectural style is derived not from a universal language but from the local context. At Amiens in 1984 he created a design which seamlessly linked the old with the new (figs. 164, 165). The logic of the old city is maintained in the new plan in three major respects: the scale of the city, the relationship of its public spaces and the typology of the local buildings.

Accepting these historical constraints, Krier has inserted a habitable bridge in the tradition of European medieval towns (see pp. 36-45).

The bridge crosses not only the canal but also streets and courtyards, creating a pedestrian axis that links the pedestrianized cathedral precinct to the north of the city (figs. 166-169). Stairs link the foot-bridge with each of the streets it crosses.

The central foot path is 4.5 metres wide, with two-storey buildings – including a hotel, a crafts centre, specialist shops, workshops and housing – running down both sides. All the buildings are faced in brick and have slate or tile roofs.

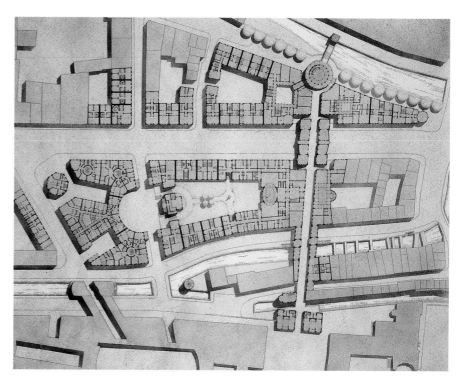

Fig. 165,
Rob Krier,
Amiens Project: site plan-II, 1984,
© Collection Rob Krier

Fig. 166, *Amiens Project: elevation*, 1984. © Collection Rob Krier

Fig. 167,
Rob Krier, *Amiens Project: view along
bridge towards Amiens Cathedral*, pencil,
1984. © Collection Rob Krier

Fig. 168, Rob Krier, *Amiens Project: view of proposed bridge*, pencil, 1984.
© Collection Rob Krier

Fig. 169, Rob Krier, *Amiens Project: view towards an inhabited bridge*, pencil, 1984.
© Collection Rob Krier

CEDRIC PRICE'S PROPOSAL
FOR HUNGERFORD BRIDGE, LONDON

While the River Thames is an exciting visual element in Central London and a useful thoroughfare for people and goods, it also creates a barrier between the two parts of the city, establishing its north and south banks as isolated entities. In 1988 the Greater London Council commissioned Cedric Price to investigate ways in which the South Bank, then largely owned by the Council, could be enhanced. Price's concept of the Solid River (fig. 170) forms part of a series of ideas arising from this commission. By creating culverts from Westminster Bridge to Waterloo Bridge and paving over the top, Price makes an uncluttered public space over eight times the size of Trafalgar Square; the open space so created would be enlarged by the incorporation of Jubilee Gardens. River transport terminals would be located at either end of the culverts. The concept recalls the ice fairs of 1683 and 1740, when the River Thames froze over; temporary market stalls (see figs. 52, 53) were erected on the ice and activities such as horse-racing and bull baiting took place.

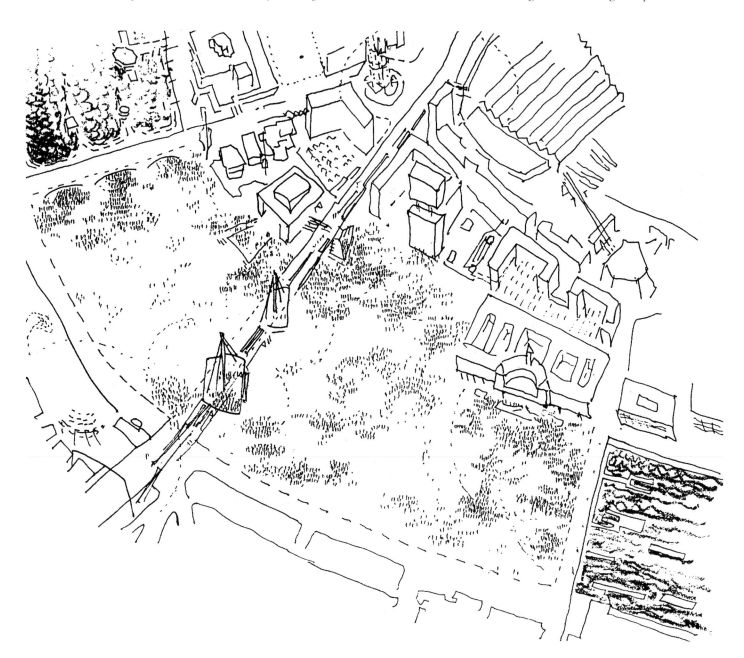

Fig. 170, Cedric Price, *Solid River: proposal for the River Thames*, pen and ink, 1988. © Collection Cedric Price

RICHARD ROGERS PARTNERSHIP'S PROPOSALS
FOR HUNGERFORD BRIDGE, LONDON

The first Hungerford bridge was designed in 1836 by Isambard Kingdom Brunel in order to attract custom to the Charing Cross market from the south side of the River Thames. It had little commercial effect; the market closed down and the land bought by the Charing Cross Railway Company who in 1864 opened the new Charing Cross Terminus. Brunel's bridge had been dismantled, its chains removed to build the Clifton Suspension Bridge (see p. 82) and its piers reused for the railway bridge with a footbridge incorporated on the downstream side. The structure provides a central London landmark notably lacking in aesthetic appeal.

Hungerford Bridge has been the subject of many rehabilitation projects. In 1986 Richard Rogers proposed its total demolition, with trains to Charing Cross Station stopping on the south side of the River Thames at Waterloo Station (fig. 171). He exhibited at the Royal Academy of Arts an island bridge to take its place complete with a transportation pod to carry passengers across the river (figs. 172, 173). The Cross River Partnership, an association of the four boroughs which front onto the River Thames in Central London, has recently organized a competition for the refurbishment of the bridge which will be part of a bid for Millennium funding.

At the Royal Academy in 1986 the Richard Rogers Partnership presented a series of urban interventions aimed at improving the links between the north and south banks of the River Thames and consolidating the public realm at the heart of London. Two radical transport interventions underpinned the proposals: the relocation of the Embankment road into a underground bypass submerged in the river bed, and the relocation of Charing Cross to form a major transport interchange with Waterloo Station.

The relocation of the road made possible the linking of all the existing gardens on the riverside into a mile-long 'linear park'. The removal of Charing Cross Station made possible the replacement of the large and ugly Hungerford Bridge with a light weight suspended pedestrian bridge supporting an undermounted shuttle train linking the South Bank Centre and its new train station with Embankment Underground Station and Trafalgar Square.

Unlike the exisiting bridges, the new one with its supporting mast was aligned with Northumberland Avenue and created a direct route from Trafalgar Square to the South Bank. In urban terms the South Bank was brought into a clear and more immediate relationship with London's most important civic space. The new bridge was designed to be as light as possible to reveal the beautiful curve of the river. Three tower elements – containing exhibition spaces, viewing platforms, cafés and restaurants – framed views and interacted with the towers of the Houses of Parliament.

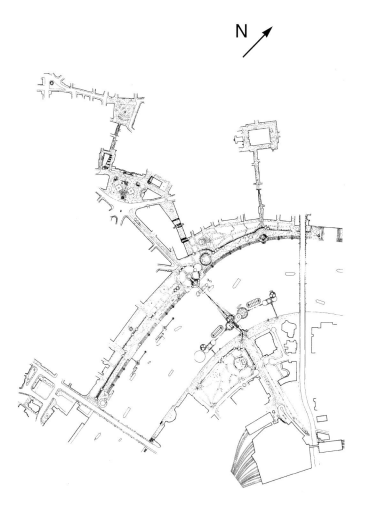

Fig. 171,
Richard Rogers Partnership,
Hungerford Bridge: London as it could be, site map, 1986.
© Collection Richard Rogers Partnership

111

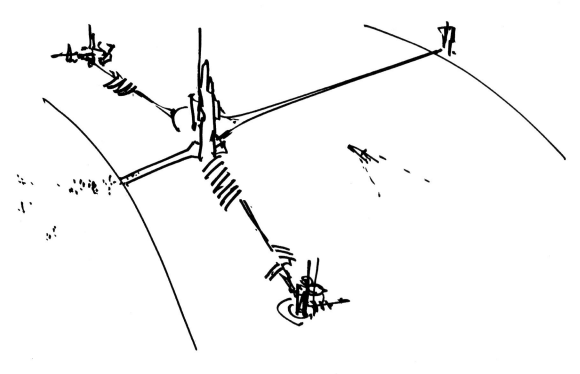

Fig. 172, Richard Rogers Partnership, *Hungerford Bridge: London as it could be, preliminary design*, pen and ink, 1986. © Collection Richard Rogers Partnership

Fig. 173, Richard Rogers Partnership, *Hungerford Bridge: London as it could be, bridge over the River Thames, model*, 1986. © Collection Richard Rogers Partnership

Fig. 174, Richard Rogers Partnership, *Hungerford Bridge: The South Bank Project, panoramic view of the bridge*, 1996. © Collection Richard Rogers Partnership

Fig. 175, Richard Rogers Partnership, *Hungerford Bridge: The South Bank Project, view of the bridge*, pen, 1996. © Collection Richard Rogers Partnership

Fig. 176, Richard Rogers Partnership, *Hungerford Bridge: The South Bank Project, view of the bridge*, 1996. © Collection Richard Rogers

Following a commission in 1995 to revitalize the South Bank Centre, the Richard Rogers Partnership pursued a range of studies aimed at improving links between the South Bank Centre and its immediate vicinity on both the north and south banks of the River Thames. The Hungerford Bridge connections were seen as a link in a chain of urban improvements aimed at creating a strong route between the Trafalgar Square and the Waterloo Station areas (fig. 174). Two routes, one local and one metropolitan, were proposed along the Hungerford Bridge axis: a travelator walkway cantilevered on the north face of the bridge linking Charing Cross Station and its locality to the South Bank (fig. 175), and a free-standing inhabited bridge along its south face forming a metropolitan main street linking Trafalgar Square to Waterloo Station (fig. 176).

Unlike the 1986 'London as it could be' proposals, the inhabited bridge accepted the constraint of the existing bridge and aimed to create a new public place suspended over the river. The effect of the new 'place' would be to draw people from both banks to the river itself and hence increase their proximity. Like its predecessor of 1986, the inhabited bridge would also contain shops, bars, cafés and restaurants, and enjoy extensive views both downstream towards St Paul's Cathedral and upstream to the Houses of Parliament.

BARFIELD AND MARKS' PROPOSAL
FOR BATTLEBRIDGE BASIN, LONDON

In the 1980s the regeneration of the run-down King's Cross area in London into an international transport interchange and office centre seemed imminent. However, delays in planning the Channel Tunnel Link to St Pancras Station and the collapse of the office market halted the proposals. Battlebridge Basin – an offshoot of Regent's Canal close to King's Cross Station – seemed at the time an ideal location for new offices. The inspired concept of building offices across the entrance to the basin not only made use of free land but also provided a building that could be a landmark in its own right (fig. 177). The bridge, whose form is heavily determined by the engineering proposal developed by Ove Arup and Partners, spans 48 metres and contains some 20,000 square-feet of office space on three floors. Steel arches provide the main support for the structure while the floors are suspended on hangers from the underside of the arch. The building is solely for office use; there is no pedestrian access for crossing the basin entrance (figs. 178, 179).

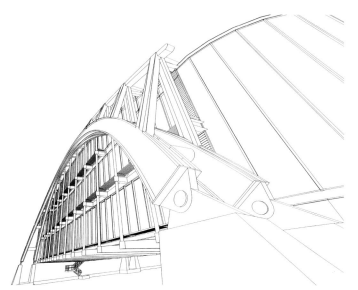

Fig. 179, Barfield and Marks, *Battlebridge Basin Bridge Building, Regent's Canal: bridge building*, 1989. © Collection Barfield and Marks

Fig. 178, Barfield and Marks, *Battlebridge Basin Bridge Building, Regent's Canal: computer rendering of structural skin*, 1989. © Collection Barfield and Marks

Fig. 180,
SITE Environmental Design,
*The Four Continents Bridge:
daytime aerial perspective*, 1989.
© Collection SITE/James Wines

SITE'S PROPOSAL FOR HIROSHIMA

This 'Garden Bridge' of 1989 in Hiroshima celebrates the links between man and the natural environment (fig. 180). Its form is based on the traditional arched bridge, which echoes a fundamental element in the Japanese landscaped garden. However, rather than simply repeating the traditional form, SITE have reinterpreted the concept by integrating contemporary technology with vegetation.

The surface of the bridge is divided by a glass wall, on one side of which are four landscaped terrariums containing vegetation from four continents. On the other side pedestrians can study a section through the layers of the earth. A vertical waterfall cascades over the glass to feed a series of streams that flow into the lake below (figs. 181-184).

Fig. 181, SITE Environmental Design, *The Four Continents Bridge: view of the bridge from the downriver side*, 1989. © Collection SITE/James Wines

Fig. 182, SITE Environmental Design, *The Four Continents Bridge: view of the entrance to the bridge*, 1989. © Collection SITE/James Wines

Fig. 183, SITE Environmental Design, *The Four Continents Bridge: vegetation cover*, 1989. © Collection SITE/James Wines

Fig. 184, SITE Environmental Design, *The Four Continents Bridge: bridge view*, 1989. © Collection SITE/James Wines

Fig. 185,
Team Luscher Switzerland, with
Jean Tonello Engineer, *Pont Devenir, Lake Geneva:
structural elevation*, 1994. © Collection Luscher

TEAM LUSCHER'S PROPOSAL FOR LAKE GENEVA

Luscher's elegant engineering solution of 1994, a 325 metre bridge which spans the western end of Lake Geneva (figs. 185-187), supports two carriageways on the outside of the structure (fig. 185) while allowing a range of activities to be accommodated within its spine.

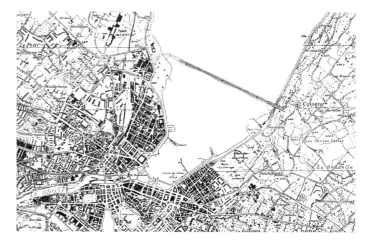

Fig. 186, Team Luscher Switzerland, with Jean Tonello Engineer,
Pont Devenir, Lake Geneva: location map, 1994. © Collection Luscher

Fig. 187,
Team Luscher Switzerland, with Jean
Tonello Engineer, *Pont Devenir,
Lake Geneva: panoramic view of the
bridge*, 1994. © Collection Luscher

116

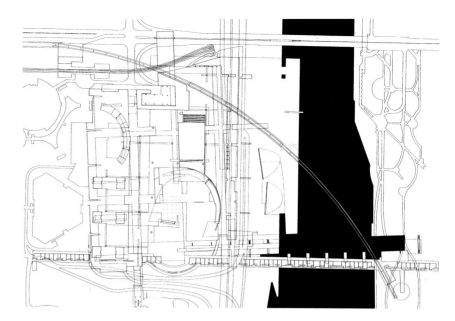

Fig. 188,
Morphosis, *Pavilion Bridge,*
Vienna Expo 1995: location map,
1995. © Collection Morphosis

MORPHOSIS' PROPOSAL FOR VIENNA

Designed in 1995 as part of an unexecuted project for the Vienna Expo on the banks of the River Danube (fig. 188), the Pavilion Bridge contains the national pavilions; it represents a literal cross-section of global culture (figs. 189, 190). There are two bridges.

One is permanent, providing service and pedestrian connections for the pavilions and the housing that would have been built on the site once the Expo was finshed; the other is a temporary structural raft that accommodates individual exhibition venues.

Fig. 189, Morphosis, *Pavilion Bridge, Vienna Expo 1995: longitudinal cross-section*, 1995. © Collection Morphosis

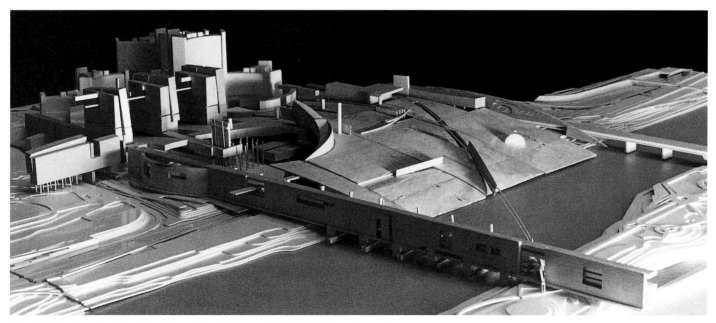

Fig. 190, Morphosis, *Pavilion Bridge, Vienna Expo 1995: model*, 1995. © Collection Morphosis

117

GEORGE PEABODY BRIDGE COMPETITION
BANKSIDE, LONDON

The Peabody Trust is the largest charitable landlord of social housing in London with over 14,000 homes, a substantial number of them in north Southwark. As part of the celebrations in 1995 to mark the bicentenary of the birth of George Peabody, the Trust held a competition for a habitable bridge spanning the Thames from the steps of St Paul's Cathedral in the City of London to Bankside – the site of the new Tate Gallery of Modern Art – in north Southwark. The Trust was eager to develop the concept of a habitable bridge 'as a microcosm of the city experience and as a model for a new urban community, with mixed residential and commercial uses, ages, income groups, high density but private homes, rented rather than owned, relatively free from noise and atmospheric pollution, carless but sufficiently central for public transport to be a real option. In short, the bridge as urban hamlet'.

As well as housing, the invited architects were asked to consider a range of uses: the University of the Third Age, an alternative site for St Bartholomew's Hospital, a National Ecumenical Centre, museums and an open-air music and performance space.

Six firms of architects presented their ideas: Hunt Thompson Associates (fig. 192), John Outram Associates (fig. 194), Levitt Bernstein Associates (fig. 191), Edward Cullinan (fig. 193), Allies and Morrison (figs. 195, 196) and Richard Horden Associates (figs. 197, 200). The architects were invited to 'collude in a conspiracy of ideas; a subtle interweaving of concepts relating to structure, urban design, relationships between cost and value, and the quality of life that should be possible in the heart of London'.

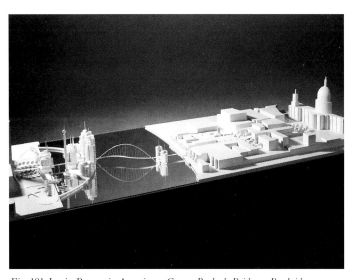

Fig. 191, Levitt Bernstein Associates, *George Peabody Bridge at Bankside: a new residential bridge for London to mark the bicentenary of the birth of George Peabody, model*, 1995. © The Peabody Trust

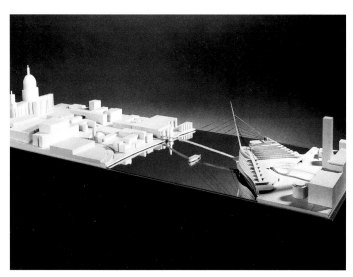

Fig. 192, Hunt Thompson Associates, *George Peabody Bridge at Bankside: a new residential bridge for London to mark the bicentenary of the birth of George Peabody, model*, 1995. © The Peabody Trust

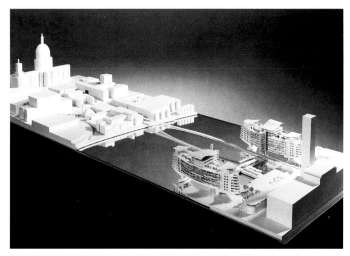

Fig. 193, Edward Cullinan and Associates, *George Peabody Bridge at Bankside: a new residential bridge for London to mark the bicentenary of the birth of George Peabody, model*, 1995. © The Peabody Trust

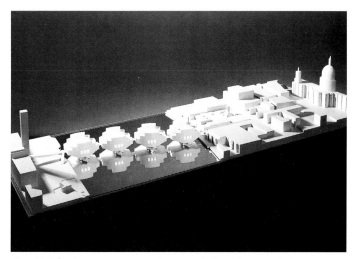

Fig. 194, John Outram Associates, *George Peabody Bridge at Bankside: a new residential bridge for London to mark the bicentenary of the birth of George Peabody, model*, 1995. © The Peabody Trust

Allies and Morrison decided to maintain the open space of the river by concentrating the accommodation into two intermediate piers (figs. 195, 196). The piers stand as islands parallel with the banks of the river, thus both breaking the structural span of the bridge and reducing the apparent length of the bridge for the pedestrian. The resulting bridge is divided into three, with a lower bridge connecting the piers to the bank and a higher open bridge crossing the central portion of the river. The proposals create two hundred housing units for the Peabody Trust.

Fig. 195,
Allies and Morrison, *George Peabody Bridge at Bankside: a new residential bridge for London to mark the bicentenary of the birth of George Peabody, competition drawing*, 1995.
© Collection Allies and Morrison

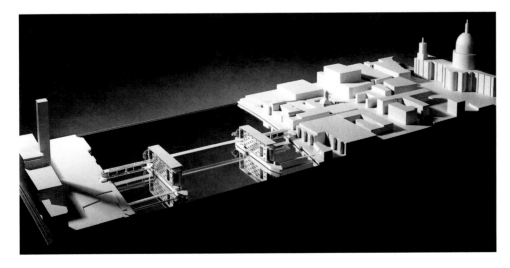

Fig. 196,
Allies and Morrison, *George Peabody Bridge at Bankside: a new residential bridge for London to mark the bicentenary of the birth of George Peabody, model*, 1995. © The Peabody Trust

119

Richard Horden's recognizably streamlined design is light in its structure and in its impact on views of the river (figs. 198, 200), the glazed structure conveying a sense of lightness and grace (fig. 197). A 200 metre gallery arcade provides a covered pedestrian route, with spectacular views of St Paul's Cathedral and other City landmarks and a central green space – a fully glazed conservatory planted with citrus trees – which creates a vantage-point for views up and down the river as well as relieving the length of the arcade (fig. 199). It is a low-energy bridge. Laminated glass photovoltaics could provide some 35 kilowatts of solar energy. Studies were undertaken to ascertain the viability of water-driven turbines. While forty per cent of the electrical power could be produced without serious environmental impact on the waterflow and habitat, the cost of maintenance of the turbines would limit the cost-effectiveness of this solution. Environmental Design was by ETH Zürich and the structural design by Mott MacDonald.

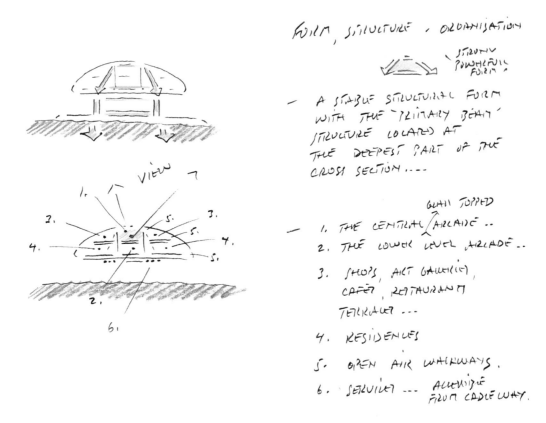

Fig. 197, Richard Horden Associates, *George Peabody Bridge at Bankside: a new residential bridge for London to mark the bicentenary of the birth of George Peabody, diagram showing form, structure and organisation of the bridge: A stable structural form with the 'primary beam' structure located at the deepest part of the cross-section ...1. The central glass-topped arcade ...2. The Lower level arcade ...3. Shops, art galleries, cafés, restaurants, terraces ...4. Residences ...5. Open-air walkways ...6. Services ... accessing from Cableway [sic].* © Collection Richard Horden Associates

Fig. 198, Richard Horden Associates, *George Peabody Bridge at Bankside: a new residential bridge for London to mark the bicentenary of the birth of George Peabody, model,* 1995. © The Peabody Trust

Fig. 199, Richard Horden Associates, *George Peabody Bridge at Bankside: a new residential bridge for London to mark the bicentenary of the birth of George Peabody, section with view of St Paul's Cathedal*, 1995.
© Collection Richard Horden Associates

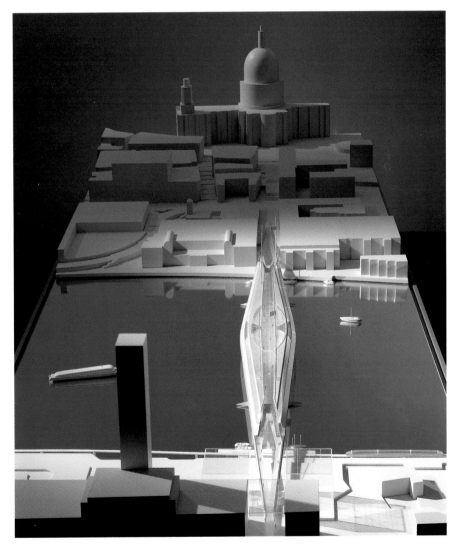

Fig. 200,
Richard Horden Associates,
*George Peabody Bridge at Bankside:
a new residential bridge for London to mark the
bicentenary of the birth of George Peabody, model*,
1995. © The Peabody Trust

BEDNARSKI'S PROPOSALS FOR ROME

Anno Santa 2000 has a special significance for Rome where it will be celebrated more as a religious festival than as the entry into a new century. Hence the significance of the Millennium Bridge designed in 1996, which lies across the River Tiber at the conjunction of two axes, the first on the Viale Angelico and Via di Porta Angelica leading to Piazza San Pietro in Vatican City, and the second extending beyond the Via Guido Reni to line up with Rome's new Mosque designed by Paolo Portoguesi (fig. 201). A bridge was proposed for the site when this quarter of Rome was first laid out between 1920 and 1940. Three streets merge where a bridge was to have been built, creating a ready-made urban context for the new structure. The bridge is designed as a flexible structure which in the first instance would house temporary living accommodation for some four million tourists and pilgrims who, it is anticipated, will visit Rome in the year 2000. Subsequently its use would be transformed to provide offices, restaurants and cafés, and exhibition and tourist facilities. A 'Garden of Paradise' runs through the centre of the bridge, representing the uniting element of Christianity and Islam, while the walls containing accommodation are pulling apart to represent the divisions between the two religions (fig. 202).

Engineers: Professor AM Michetti with Dewhurst Macfarlane and Partners.

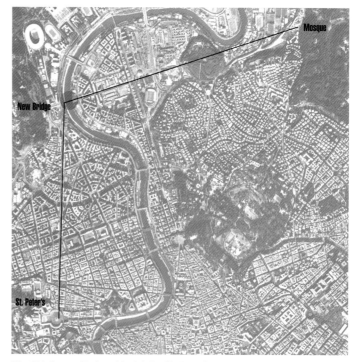

Fig. 201, Cezary Bednarski with Studio E Architects, *Millenium Bridge, Rome: location map*, 1996

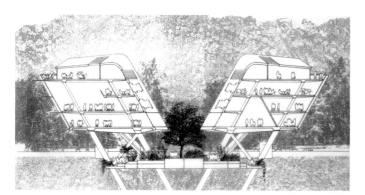

Fig. 202, Cezary Bednarski with Studio E Architects, *Millenium Bridge, Rome: cross-section*, 1996. © Collection Cezary Bednarski

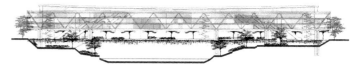

Fig. 203, Cezary Bednarski with Studio E Architects, *Millenium Bridge: longitudinal cross-section*. © Collection Cezary Bednarski

The project Acrobeleno 2000 proposes twenty temporary hotel bridges, providing 2,000 rooms (4,000 bed spaces) to ease the severe accommodation shortage forecast for the year 2000 (fig. 204). The advantage of building over the River Tiber is that there will be no disruption of archaeological remains – a major deterrent for development in the city. High-tech room modules would be leased for the year and slotted into the modular structure. The hotels would provide rooms only, with reception areas located at each end. Guests would use local restaurants and bars for their catering requirements. Each bridge would be painted a different colour based on the hues of the rainbow.

Engineer: Jane W Wernick, Ove Arup and Partners.

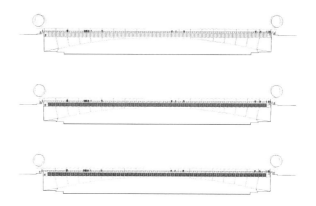

Fig. 204, Cezary Bednarski with Studio E Architects, *Progetto Arcobeleno 2000/ Millenium Bridge: elevations of the proposed hotel-bridges*, 1996. © Collection Cezary Bednarski

Fig. 205,
Mario Bellini Associati,
*Pearl Bridge, Dubai:
preliminary design*, 1996.
© Collection Mario
Bellini Associati

BELLINI'S PROPOSAL FOR DUBAI

Dubai is a divided city, with Deira lying to the north of the Al-Khor Creek and Bur Dubai to the south. The only connections between the two sides are a tunnel and two vehicular bridges. The Dubai Pearl Bridge is intended to weld the two sides together to create a more cohesive city (figs. 205, 206).

Abu Dhabi is rapidly becoming the financial centre of the Middle East and recognizes the need for a strong landmark building which could include a conference centre of international standard. Bellini's design of 1966 incorporates a 2,500-seat auditorium in the shape of a pearl (a reference to the pearl-diving industry for which Adu Dhabi is traditionally famous) forming the keystone of the structure. Surrounding the auditorium are 200,000 square metres of usable floor space which incorporates facilities related to the conference centre, the United Arab Emirates' Stock Exchange, a 400-room hotel, offices, business suites, clinics, entertainment venues and luxury apartments. A fourteen-metre-wide gallery spans the creek within the building, while an external passerella or foot-bridge, provides a comfortable, palm-shaded promenade for use during the mild season. The structure is 480 metres long, 90 metres high and 60 metres wide. At both ends of the bridge there will be large car parks to provide easy access to the bridge facilities and to reduce the need for vehicles to cross from the south into Deira, which is already heavily congested.

Architect: Mario Bellini Associati
Structural Consultants: Redsco/Giuliani.

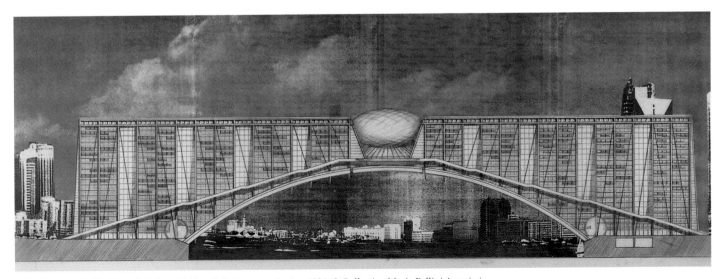

Fig. 206, Mario Bellini Associati, *Pearl Bridge, Dubai: panoramic view*, 1996. © Collection Mario Bellini Associati

REITER'S PROPOSAL FOR BOSTON

The Old Northern Avenue Bridge was built in 1908 as a swing bridge providing a vehicular link between the two industrial waterfronts flanking Boston's Fort Point Channel (fig. 207). Today it links the thriving downtown waterfront area and financial district to Fort Point, the district identified for Boston's next phase of urban development, with proposals for a new convention centre, office buildings, housing, a waterfront project and facilities for water transportation. The City of Boston is unwilling to provide for the maintenance of the existing bridge and has stipulated that the bridge must remain permanently fixed in its open position in order to maintain a navigational passage. In 1966, the Boston Redevelopment Authority invited Wellington Reiter to make proposals that would preserve the bridge as an essential link with the new development area. His proposals include locking the bridge into an open position with additional supports, one at either end. This central island would then be accessed by a series of smaller bridges (figs. 210, 211).

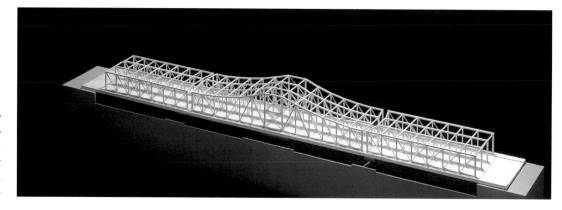

Fig. 208,
Wellington Reiter,
Northern Avenue Bridge, Boston,
Massachusetts: model showing the
swing bridge closed, 1996.
© Collection Wellington Reiter

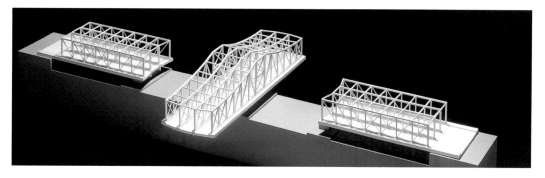

Fig. 209,
Wellington Reiter,
Northern Avenue Bridge, Boston
Massachusetts: model showing the
swing bridge open, 1996.
© Collection Wellington Reiter

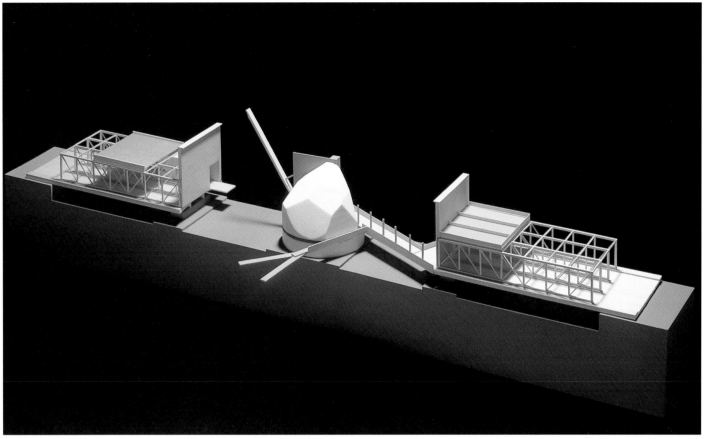

Fig. 210, Wellington Reiter, *Northern Avenue Bridge, Boston, Massachusetts: model*, 1996. © Collection Wellington Reiter

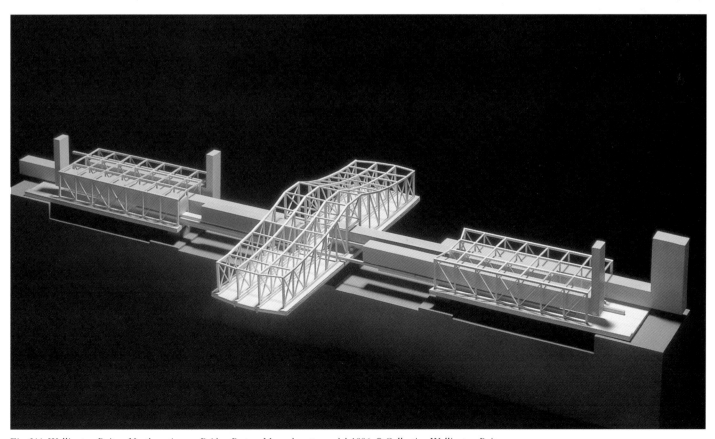

Fig. 211, Wellington Reiter, *Northern Avenue Bridge, Boston, Massachusetts: model*, 1996. © Collection Wellington Reiter

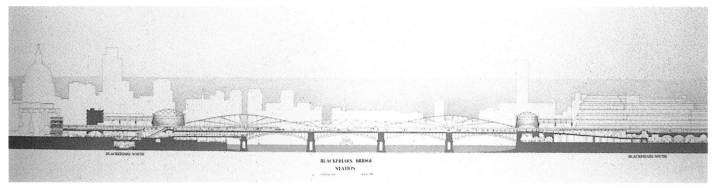

Fig. 212, Terry Farrell and Partners, *Thameslink 2000: Blackfriars Bridge Station: longitudinal elevation*, 1991. © Collection Terry Farrell and Partners

TERRY FARRELL'S PROPOSAL FOR BLACKFRIARS BRIDGE STATION, LONDON

Farrell's double-ended station of 1991 creates a new point of connection between the City in the north and the more isolated South Bank (fig. 212). To the north, Farrell proposed a station hall within the existing buildings and, to the south, a new station building above an existing car-park. The requirements of the new line and the geometry of the north bank station meant that the Thameslink platforms had to be located on the bridge itself; the listed bridge was then covered with a curvilinear superstructure, which also contains a travelator to facilitate movement across the river and a grand concourse running beneath the platforms (figs. 213, 214).

Fig. 213, Terry Farrell and Partners, *Thameslink 2000: Blackfriars Bridge Station: model*, 1991. © Collection Terry Farrell and Partners

Fig. 214, Terry Farrell and Partners, *Thameslink 2000: Blackfriars Bridge Station: model*, 1991. © Collection Terry Farrell and Partners

Fig. 215,
Alsop and Störmer,
Blackfriars Bridge: model, 1995.
© Collection Alsop and Störmer

ALSOP AND STÖRMER'S PROPOSALS FOR BLACKFRIARS BRIDGE, LONDON

Blackfriars is an important hub lying to the west of the City of London. It boasts a road bridge, a railway bridge and the remaining piers of the Dover, Chatham and London Railway, from which the rails were removed in the 1980s. The extant railway bridge forms the river crossing of the Thameslink 2000, which involves upgrading the capital's north-south rail links to provide direct services from Brighton to Bedford.

Alsop and Störmer were initially engaged to consider the possibility of using the surviving piers of the DCL railway bridge as the support for a deck which could provide a new location for the Institute of Contemporary Arts (fig. 218). The practice was subsequently commissioned to extend the area of their work to incorporate the renovation of the existing Thameslink railway bridge. Alsop and Störmer proposed to throw over the bridge a lightweight, transparent station roof, and to provide accommodation for station facilities (figs. 215-217).

The design includes a new, high-level walkway, which will offer enhanced views of the River Thames and St. Paul's Cathedral.

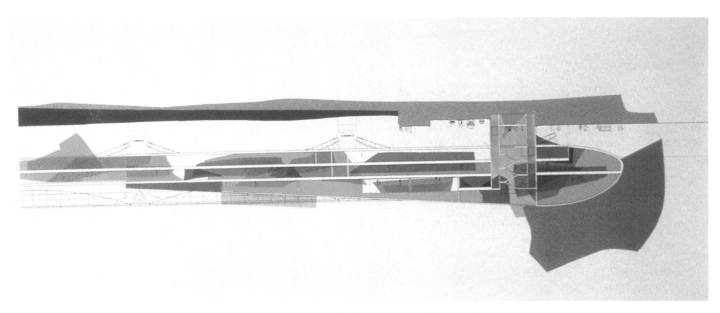

Fig. 216, Alsop and Störmer, *Blackfriars Bridge – a new site for the ICA: longitudinal section*, 1995. © Collection Alsop and Störmer

Fig. 217, Alsop and Störmer, *Blackfriars Bridge – a new site for the ICA: cross-section*, 1995. © Collection Alsop and Störmer

Fig. 218, Alsop and Störmer, *Blackfriars Bridge – a new site for the ICA: photo-montage*, 1995. © Collection Alsop and Störmer

Thames Water Habitable Bridge Competition

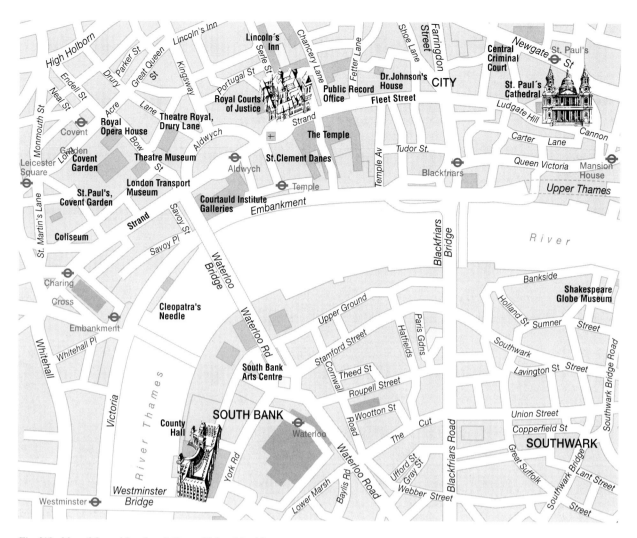

Fig. 219, *Map of Central London*. © Franz Huber, Munich

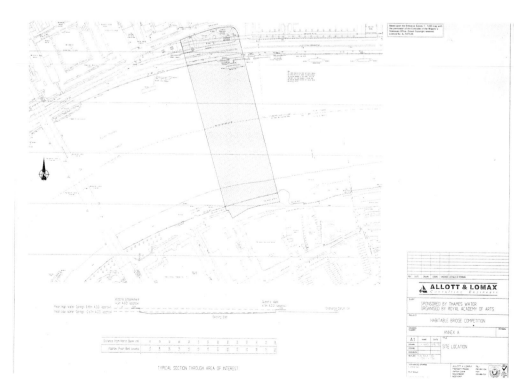

Fig. 220,
*Thames Water Habitable Bridge
Competition: site plan*, 1996. © Allott
and Lomax Consulting Engineers

THE VIABILITY OF THE INHABITED BRIDGE IN THE TWENTY-FIRST CENTURY

Stuart Lipton and David Cadman

Within the context of the exhibition *Living Bridges: The inhabited bridge, past, present and future*, the Royal Academy of Arts held an international competition to design an inhabited bridge across the River Thames, between Temple Underground station on the north bank and the London Television Centre on the south. The competition was intended to contribute to the current debate about the use of the river, the accessibility to areas on both banks and to urban planning in general. Seven firms of international architects were asked to submit detailed design proposals which would then form the climax of the exhibition. The design brief for the competition made it clear that the bridge should be seen as a destination in its own right; should be of mixed use, i.e. residential, commercial and cultural; should be for pedestrians only, and should maintain a firm foothold in reality in terms of its usage, structure and materials. Most importantly it should be cognizant both of navigational and other requirements imposed by the Port of London Authority and demonstrate an awareness of its relationship to the surrounding city.

The design brief also demanded that the bridge should have the potential to be self-funding. With this in mind, the competition project was the subject of a feasibility study undertaken by the international management consultants KPMG, who were asked to review the commercial viability of the bridge and to identify the factors which would underpin that viability. Their report, which confirmed that a bridge could be commercially successful, formed part of the design brief for the competition. It was hoped that the competition, and the exhibition with its review of the history of the inhabited bridge as a building type, would enhance the possibility of realizing the bridge.

The conventional approach to any consideration of the viability of an inhabited bridge is to focus primarily upon the physical and financial feasibility of the task: can it be built and where will the money come from to pay for it? But these two questions are in reality only supplementary questions. The primary question concerns the function of the bridge and whether or not it will 'come alive'. All too often, the feasibility of schemes of urban development and revitalization become bogged down in spreadsheets of cost and return and, from the outset, too little attention is given to what will happen there, the sense of place, the *genius loci*. Thus, individual buildings and even major schemes of development have been built primarily to meet the perceived needs of the investor rather than those of the occupier, and some of the recent calamities in London and, indeed, in other cities, both in the United Kingdom and elsewhere, are evidence of this. It would appear, therefore, that the questions should be reversed. Firstly, how will people use this bridge and what kind of experience might it offer? Secondly, if the bridge is to 'come alive', how must it be designed and constructed and is this possible? And then, thirdly, given its use and construction, can it be viable in terms of costs and returns?

The essence of any bridge is that it must be well founded on each bank. It must lead from somewhere to somewhere. It cannot be a solitary construction whose task is merely to span the river. It must fit into the urban fabric of which it is part. In this particular case, the fabric is extensive, for the bridge would be part of a network of present and proposed bridges across the River Thames and would form part of a pathway linking important buildings and places on the north and south banks of the river. For example (see fig. 219), the pathway might take the pedestrian from Covent Garden, past the new Opera House, along the Strand, through Somerset House to the new Terrace, across the bridge to the South Bank Centre, to the Globe and the new Tate at Bankside and to Tower Bridge, back across the river to the Tower and then to St Paul's Cathedral.

However, although the bridge has to be seen as part of a larger plan, it would also, in a sense, redefine the context of that plan. Its presence would alter the significance of what is already there, and would introduce into the plan questions of civic intention and integration – a deliberate act of recognizing that the city as a whole is more than the mere sum of its parts. At the same time, the bridge is not to be a rudimentary pedestrian bridge, merely linking one side of the river to the other. It is to be a destination in its own right, an inhabited bridge.

The uses discussed in the report prepared by KPMG included office, residential, retail and recreational/educational, favouring a mix of retail, restaurant and residential. However, there is here an important difference of quality that has to be considered. Residential is essentially a private and self-contained use, while retail, restaurants and, indeed, recreational/educational uses are essentially public and accessible. Given that the bridge is to be a part of a lively pathway, the emphasis should therefore be upon retail/restaurant and recreation/education. Some of these uses will be highly profitable and these will help to support those that are not. In any event, the full mix and balance of uses is essential. One can imagine coming on to the bridge from the terrace of Somerset House, browsing in shops and perhaps stopping for coffee and cake in a café overlooking an open square at the centre, with fine views of St. Paul's and the Houses of Parliament. You might continue your walk, stopping at a museum or gallery celebrating London and its great river. You may be a visitor, you might come here for lunch or agree to meet with friends in the evening after work. The bridge would welcome and invite you across.

Finally, in terms of use, there is the practical question of the management of the bridge and, in particular, of the public realm, the spaces between the buildings. This is not a major problem for it is a familiar task. There already exists considerable experience of such management, for example in large shopping centres, office campuses and new office developments such as Broadgate in the City of London. The cost of such management and the services that are provided are normally recovered by way of a service charge paid by the organisations occupying the buildings, and there is a growing awareness of the need for occupiers to cooperate with each other – for example in the recent formation of the Association of Town Centre Management which now has managers operating in over 170 towns and cities in the UK. This experience recognizes the fact that successful places are those where attention is given to the public as well as to the private realm and that this requires a working partnership between the public and private sectors, the government of the city, its traders and its people.

There are, of course, constraints upon the form and use of the bridge and, in particular, upon the configuration of the buildings. Not only have the requirements of the Port of London Authority to be considered but, more importantly, consideration has to be given to the sight-lines of the views of St. Paul's Cathedral and the Houses of Parliament. Does the bridge obstruct or enhance these views? By bringing people to the centre of the river, might it not create one of the most wonderful and dramatic river views in London? Another constraint is the need to provide access for public services such as fire, ambulance and refuse collection, which would be particularly stringent if office and residential uses were to be included. Nevertheless, the KPMG report suggested that these needs can be accommodated.

The KPMG report did not suggest that there were any insuperable problems to the construction of the bridge. Two particular matters, however, require especial thought. Firstly, and practically, there is the question of the way in which the ends of the bridge spring from each bank. To make the bridge accessible and 'alive', these points of contact need to encourage passage. On the north side, the problems include the height of the bank, the location of the Temple Underground station and the possible links with the new Somerset House Terrace. On the south, the problem is about the treatment of the area in front of the London Television Centre and this area's present lack of identity. But the bridge provides an opportunity for enlivening the area. Secondly, there is the question of the built form of the bridge. The bridge may break rules and convention and, indeed, may set a precedent. This can only be justified if the bridge not only forms an integral part of a wider vision for London and demonstrates substantial public benefit but also is, in its own right, a masterpiece of architecture and engineering. Not a mere bridge with some shops on it but a construction that is marvellous and inspiring, from its foundations to its topmost point. The bridge would be part of a great tradition of inhabited bridges, a tradition that includes the old London Bridge which, in the thirteenth century was a gate-way to the city and provided houses and the chapel of St Thomas. No doubt the challenge for the designers is to respond both to this tradition and to the present and future celebration of London and the Thames.

It is here that the question of financial viability must be addressed. The KPMG report looked at this in some detail and concluded that the scheme 'should certainly be viable'. This should come as no surprise, for the bridge represents a prime site in the centre of the city with quite exceptional views. Nevertheless, investors would need to be persuaded that the bridge has value in their terms. Once again, attention should be drawn to where that value comes from. It does not come from the developer's spreadsheet and calculator, nor indeed, merely from physical design. It comes from the extent to which a place has been created, a place that people are attracted to, a place that 'comes alive'. In particular, in this case, investors will need to be shown that the uses proposed, the linkages with each bank and the flow of people to, from and across, will generate a satisfactory stream of income. The success of London's Covent Garden and similar schemes in other cities demonstrate that this can be done. Can it be done here?

The inhabited and 'accessible' bridge that we have in mind will primarily comprise retail/restaurant and recreational/educational uses. In the case of the retail/restaurant space, the rents paid by the traders will form the necessary stream of income for which investors can be expected to pay a capital sum. Both domestic and foreign investors are likely to be attracted to such an investment project. The arithmetic of capital value is not complicated and KPMG's initial analysis shows that projected levels of rent and investment yield are favourable. They suggest that the capital sums generated for such a scheme would be likely to cover the probable cost of the bridge and might well create a surplus that would go towards helping to fund elements of the scheme such as museums and galleries that may not be self-funding, no doubt in partnership with other civic initiatives.

The success of this bridge will rest upon the place that is created. The underlying principles are of integration and coordination, of recognizing relationships. Success will depend upon turning convention on its head. Instead of starting with the physical and the financial and hoping that activity will follow, one should start with activity, for it is upon this that all else rests. Imagine a piazza on the River Thames, the chatter and bustle, the views eastward and westward, the traders, visitors and people working nearby. Start from this living and working community and then the physical and the financial can play their part. Successful communities mean successful businesses and successful businesses help to create successful cities.

INTRODUCTION TO THE COMPETITION

Peter Murray

The Thames Water Habitable Bridge Competition was organised by the Royal Academy with the support of the Government Office for London. Its purpose was to generate discussion about the feasibility of creating a new inhabited bridge over the River Thames in London.

The Thames Strategic Study, published in 1995, identified two sites on the River Thames where a new bridge would improve communications between the two banks. The one selected for the competition lay within a 50 metre wide band and spanned the river from Temple Gardens in the north, adjacent to Temple Underground station, to the London Television Centre on the south. The brief called for a bridge which would reduce the divisive nature of the River Thames, become a destination in its own right and not be seen in isolation from the rest of the city. Furthermore, the bridge should be buildable, structurally sound, commercially viable and capable of meeting the stringent demands of a navigable span of 160 metres imposed by the Port of London Authority, the body which controls navigation on the River Thames, in order to ensure visibility for craft negotiating the riverbend between Waterloo Bridge and Blackfriars Bridge.

A feasibility study, undertaken by KPMG, formed part of the brief. It demonstrated that the construction of an inhabited bridge could be funded through revenue generated from its lettable space. This was calculated to range from a minimum of 15,000 square metres to a maximum of 45,000 square metres, although it was felt that this latter was likely to produce a structure whose volume would mean that it might lie beyond acceptable aesthetic limits.

Seven architects of international standing were invited to enter the limited competition. The practices were selected because they represented a broad spectrum of approaches to urban problems and were therefore judged likely to bring fresh thinking to bear on the revived typology of the inhabited bridge. The judges were not disappointed. The entrants literally tore apart the concept of a bridge and produced a range of designs that will certainly inform the ensuing debate.

The judges selected two designs: the elegant deconstructivist inhabited bridge by Zaha Hadid Architects which cantilevered shards of buildings into the centre of the river, and the more traditional urban approach of Antoine Grumbach which proposed a pedestrian walkway flanked by flexible retail and café units springing from the base of two landmark towers. Most of the submitted designs confirmed the view of the KPMG feasibility study that sufficient accommodation can be included within the bridge to fund its construction.

A major factor in the acceptance of a proposal for an inhabited bridge across the River Thames is that of the reaction of the public. All the designs presented for the competition are the subject of a poll among visitors to the Royal Academy and are available for inspection by developers whose interest is vital if London is once again to enjoy the contribution that an inhabited bridge could make to the quality of its urban life and its environment.

Fig. 221, *Thames Water Habitable Bridge Competition: view of the site from Waterloo Bridge*, 1996. Photograph, Simon Hazelgrove

ZAHA HADID

The bridge is arranged as a series of cantilevered volumes linked in the centre by light pedestrian walkways. Public activities take place on the lower levels of the bridge and private accommodation is contained within the five separate building volumes above. Flexible, multi-functional loft spaces are designed to be used as residential and office spaces, artist studios and workshops. They are constructed within the space of the structural trusses. Each truss forms one building volume physically separated from each other, so that all the lofts are naturally lit and ventilated.

The trusses are lifted high above the water to allow the formation of the suspended public space below. The bridge would be open 24 hours a day and accommodate a mixture of commercial, cultural, entertainment and recreation functions.

Architectural Design	Office of Zaha Hadid, Zaha M Hadid
Competition Team	Patrik Schumacher, Graham Modlen, Ljiljana Blagojevic, Paul Karakusevic, Woody Yao, Markus Dochantschi, Tilman Shall, Thilo Fuchs, Colin Harris, Shumon Basar, Katrin Kalden, Anne-Marie Foster
Models	Alan Houston, Michael Howe
Computer Design	Wassim Halabi, Simon Yu, Garin O'Aivazian
Structural Engineer	Ove Arup and Partners, Jane Wernick, Sophie La Bourva
Services Consultant	Ove Arup and Partners, Simon Hancock, Dorte Rich Jorgensen
Transportation Consultant	Ove Arup and Partners John Shaw
Management	Ove Arup and Partners (PMS) Harry Saradjian
Cost Consultant	Davis Langdon & Everest Rob Smith, Sam Mackenzie

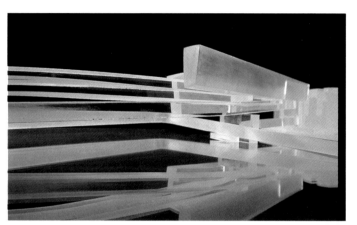

Fig. 222, *Model.* © Collection Zaha Hadid

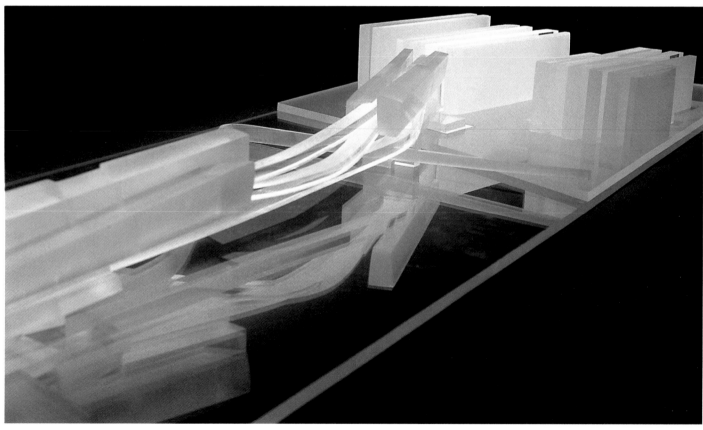

Fig. 223, *Model.* © Collection Zaha Hadid

Fig. 224, *Computer rendering of proposed bridge from the South Bank*. © Collection Zaha Hadid

Fig. 225, *Preliminary drawing*. © Collection Zaha Hadid

Fig. 226, *Preliminary drawing*. © Collection Zaha Hadid

Fig. 227, *Preliminary drawing*. © Collection Zaha Hadid

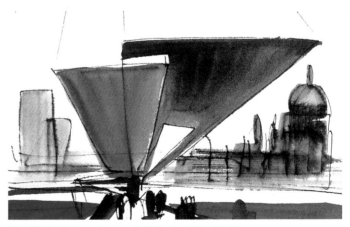

Fig. 228, *Preliminary drawing*. © Collection Zaha Hadid

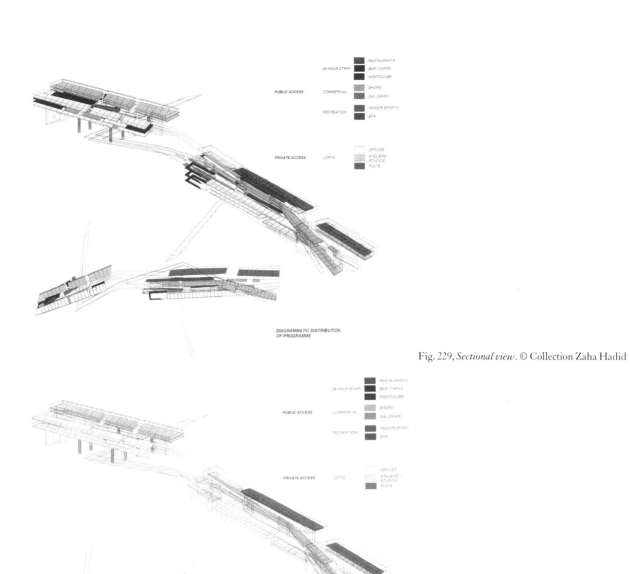

Fig. 229, *Sectional view*. © Collection Zaha Hadid

Fig. 230, *Sectional view*. © Collection Zaha Hadid

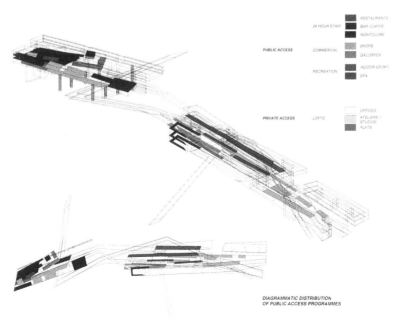

Fig. 231, *Sectional view*. © Collection Zaha Hadid

138

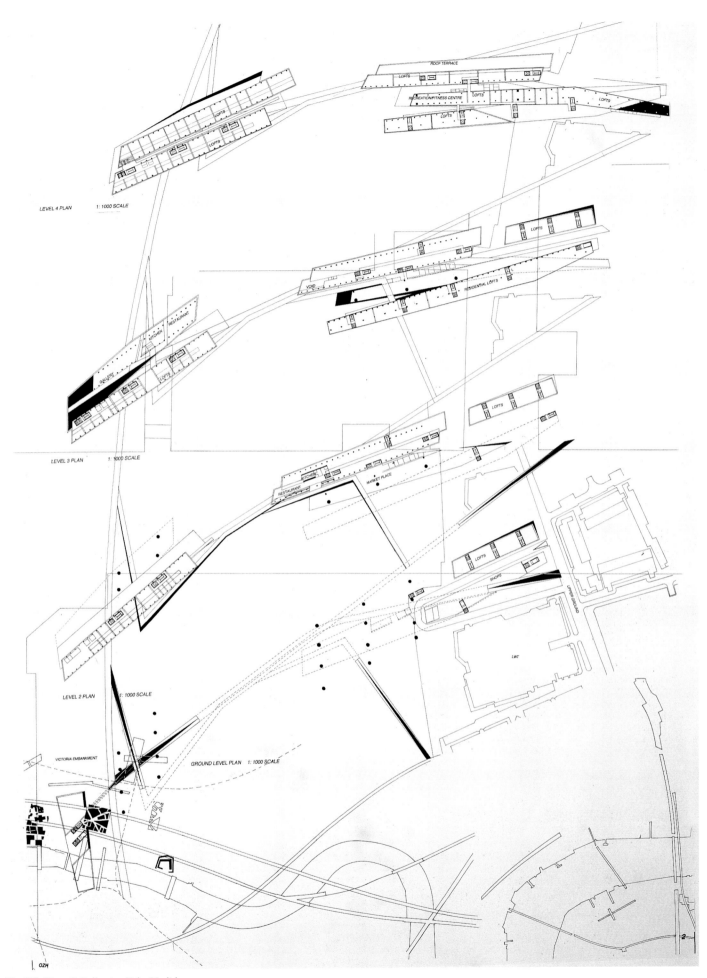

LEVEL 4 PLAN   1: 1000 SCALE

LEVEL 3 PLAN   1: 1000 SCALE

LEVEL 2 PLAN   1: 1000 SCALE

GROUND LEVEL PLAN   1: 1000 SCALE

Fig. 232, *Plans*. © Collection Zaha Hadid

ANTOINE GRUMBACH: THE GARDEN BRIDGE

In the bend of the River Thames, the 'Garden Bridge' links the two banks of the river with a series of gardens placed on either side of a covered arcade. The 'Garden Bridge' is conceived as a structure able to accommodate a variety of functions which can change over time.

Architect	Antoine Grumbach
Engineer	Marc Mimram
with	Celine Carton, Thibault
	Thierry Bruchet, Olivier Boesch
Landscape Architect	Lena Soffer

The Garden Bridge consists of three elements. On the south side, the 'World's Culture Greenhouse' is a vast covered public space protecting plants and tropical trees and providing space for restaurants, shops and flexible spaces for live concerts and other public activities. Access to the level of the bridge is gained by interior and exterior staircases, lifts and escalators. At water level, on either side of the green house, there are two 'board walks' which provide links between the bridge and the river bank. The Garden Arcade lies between the greenhouse and the towers. Hedges running at right angles to the bridge's axis provide divisions between the shops and restaurants situated on the bridge. Three sections open directly onto the river, to allow the river to be crossed in the open air. The Hanging Gardens Towers which support the cables for the suspended portion of the bridge contain a hotel and apartments, with restaurants and meeting spaces conceived as greenhouses within and at the top of the towers. Their façades are covered with a double metal skin which supports the vertical gardens. The crowns of the towers are adorned with two winged shapes which evoke flight towards the 3rd Millenium.

Fig. 233, *Model*. © Collection Antoine Grumbach

Fig. 234, *Aerial perspective*. © Collection Antoine Grumbach

Fig. 235, *View from Waterloo Bridge at night: montage*. © Collection Antoine Grumbach

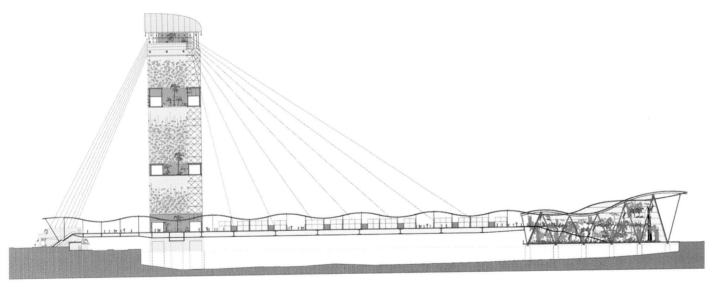

Fig. 236, *Longitudinal section*. © Collection Antoine Grumbach

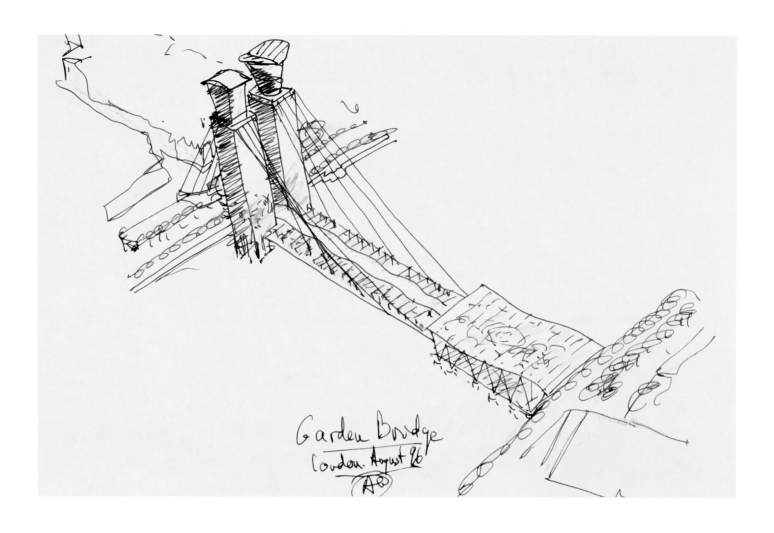

Garden Bridge
London. August 86
AG

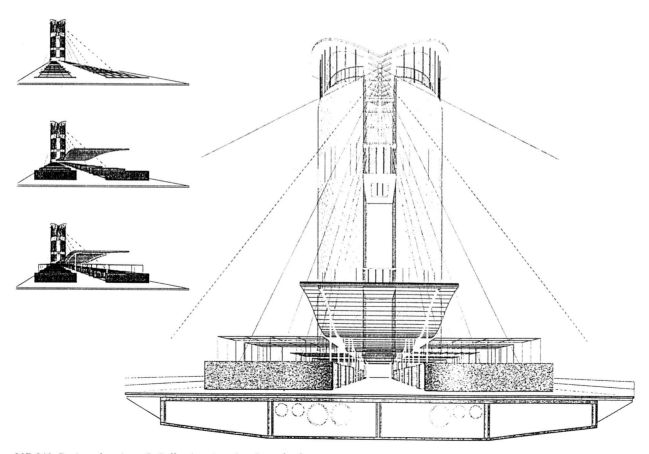

Figs. 237-240, *Projects drawings*. © Collection Antoine Grumbach

142

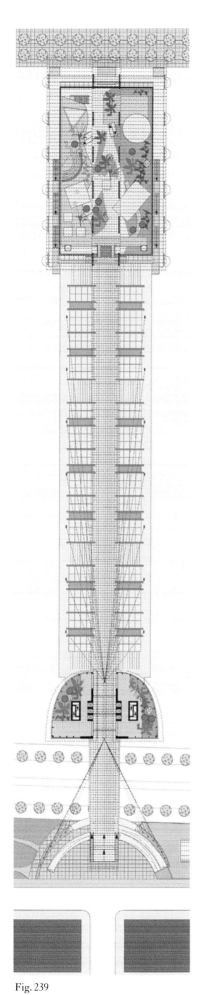

Fig. 239

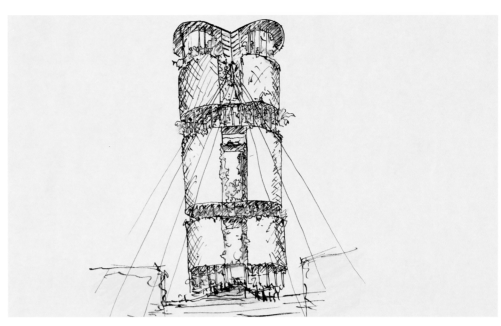

Fig. 240 (a)

Fig. 240 (b)

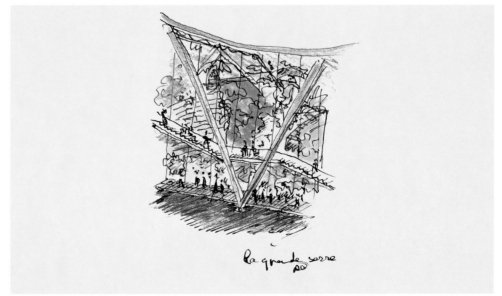

Fig. 240 (c)

143

BRANSON COATES: BRIDGE CITY

Bridge City will provide an expo-scale destination, open twenty-four hours a day and 365 days a year, incorporating an hotel, retail areas, restaurants, auditoria, malls and entertainment. The bridge is aligned with Arundel Street to the north, taking up an axial position that connects the bridge to the Strand. To the south the towers of the hotel coincide with the London Television Centre to form an effective gateway to the bridge and the north bank.

The project consists of a pair of towers, which form the southern-most pier, and a sheathed, two-storey structure with a publicly accessible roof spanning between the two pier points. Externally the building is intended to have an exquisite organic quality, the net-tensioned solid skin is coated with an iridescent surface that changes colour with the angle of view and light. At night the building will glow and shimmer.

Varied routes alter the relationship between the inside and the outside of the building. Approaching from the north bank, visitors will start their visit in relatively enclosed space, where the first shops and auditorium are housed. These routes expand outwards towards the centre of the River Thames, where huge ocular windows look out over the riverscape. On the restaurant level, a balcony extends the dining areas out towards the water.

Approaching from the south bank, where visitors have closer access to the water's edge than on the north, a promenade route extends from Westminster Bridge, past the South Bank Centre, towards the bridge. Here the design exploits the proximity to the water by first leading the visitor towards a giant pontoon that moves up and down the piers of the hotel with the tide.

Architect	Branson Coates
Design Team	Nigel Coates, Doug Branson,
	Carlos Villanueva Bradt,
	Gemma Collins, Norman Roberts,
	Gerrard O'Carroll, Geoffrey Makstutis,
	Christian Ducker, Allan Bell
Consultants	Sir Terence Conran, Ben Evans
Structural Engineers	Mike Cook,
	Chris Williams at Buro Happold
Quantity Surveyors	Colin Scattergood at D.L & E

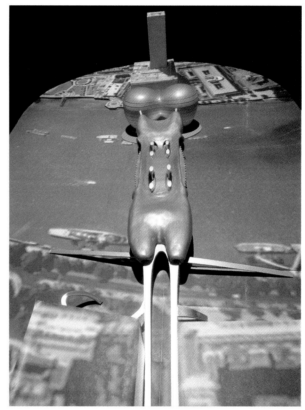

Fig. 241, *Model*. © Collection Branson Coates

Fig. 242, *Diagrammatic drawing*. © Collection Branson Coates

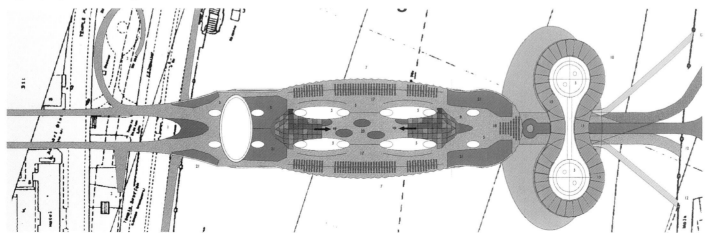

144

Fig. 243, *View from Waterloo Bridge: montage*.
© Collection Branson Coates

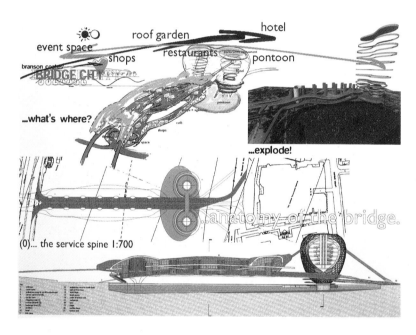

Fig. 244, *Diagrammatic drawing*.
© Collection Branson Coates

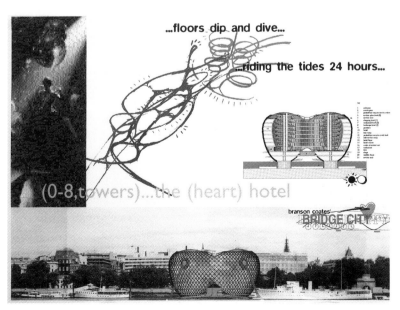

Fig. 245, *Diagrammatic drawing*.
© Collection Branson Coates

FUTURE SYSTEMS: THE PEOPLE'S BRIDGE

The architects conceived the bridge as 'a fluid and organic form sympathetic to the natural forces of the river tides. The introduction of colour to the skin animates the bridge adding visual interest and accentuating the slender elegant curves of the form. The soft edges of the elliptical structure minimise the building mass over the water'.

There are two levels of circulation: shops, stalls, kiosks and bars are located at the main deck level as are the stepped seating areas that offer panoramic views of London. The loose fit planning of the deck is designed to encourage spontaneous events and happenings as well as occasional markets and fairs. The top deck, constructed of glass, creates an open air route across the river. In the centre of the bridge the main deck passes through the body of the structure giving spectacular views down onto the restaurant and the activities on the bridge, as well as giving access to the lower decks.

The bridge is formed from a continuous double-skinned semi-monocoque structure of thin steel plates, joined together using ship-building technology. The lightweight central suspended span is carried by heavier, cantilevered side sections spanning from the principal piers.

Architects	Future systems
Structural Engineer	Techniker, Matthew Wells
Environmental Engineer	Ove Arup & Partners
	Andrew Sedgwick
Quality Surveyor	Hanscombs, Jonathan Harper
Urban Strategy	Richard Burdett
	Director, City Policy,
	Architecture & Engineering,
	London School Of Economics
Planning Strategy	Eric Sorensen Chief Executive
	London Docklands
	Development Corporation
Development Strategy	Ken Dytor
Construction Advice	British Steel, Robert Latter MBE
Landscape Architects	Townsend Landscape Architects

Fig. 246, *Model*. © Collection Future Systems

Fig. 247, *View from Waterloo Bridge: montage*.© Collection Future Systems

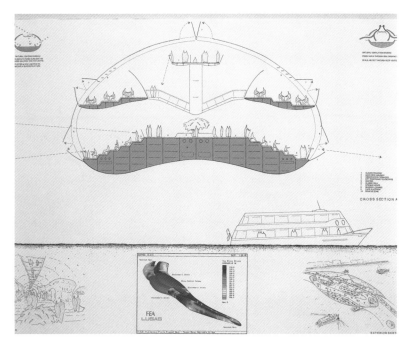

Fig. 248, *Cross-section*.
© Collection Future Systems

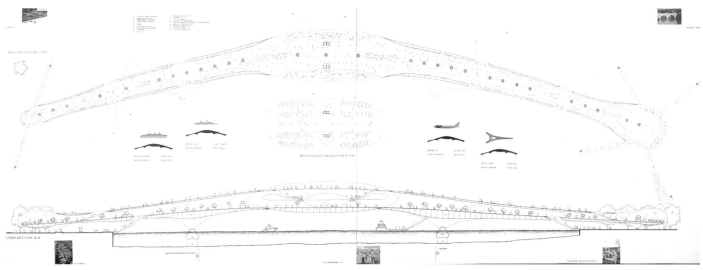

Fig. 249, *Longitudinal cross-section*. © Collection Future Systems

147

KRIER KOHL

The proposed bridge is designed to integrate the lessons of the past with the knowledge of the present, and develop an urban framework which reinforces rather than fragments the fabric of the city.

Most of the habitable area is devoted to an hotel, office space, retail space and residential units. In order to meet the Port of London Authority's requirement for a 160 metre clear span, the majority of the functions have been concentrated in the two towers and the gate houses on each bank of the river.

The main floor of retail space runs the length of the bridge, with restaurants and cafés around the central court. The central span of the bridge is a visually lightweight structure of steel and glass, three floors at the centre and rising to a point between the main towers. The two floors above the main level between the towers are also devoted to retail and entertainment, while the floors to the north are offices and those to the south are wings of the hotel.

The two towers of residential units rise ten floors above the main level, with views up and down the river. The average size of the apartments would be some 150 square metres, with more spacious penthouses on the upper floors.

The office space is concentrated on the north bank, in the three gate-houses which mark the entrance to the bridge and which frame the two flights of stairs leading to the main level. The corresponding gate houses on the south bank house the hotel.

Architects	Rob Krier,
	Christoph Kohl
Engineers and Planners	Arup GmbH Berlin
Quantity Surveyors and Cost Consultants	Davis Langdon
	& Weiss
Development Consultants	MAB Groep B.V.

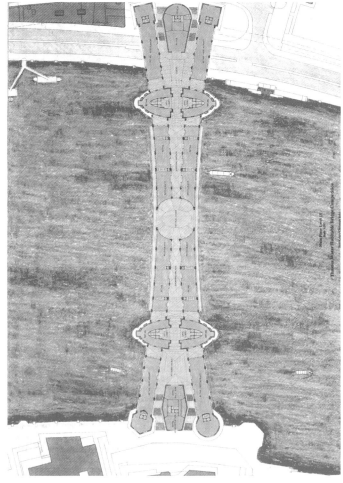

Fig. 250, *Plan*. © Collection Krier Kohl

Fig. 251, *View from Waterloo Bridge: montage*. © Collection Krier Kohl

Figs. 252-255, *Perspective drawings*. © Collection Krier Kohl

149

ARCHITECTBÜRO LIBESKIND: X-WEB WITH A POINT

Libeskind has rethought the idea of a bridge for the 21st century with a structure that celebrates the River Thames as the centre of London through a multivalent, non-linear system of connections.

The proposal includes two components. The first is a vertical habitable bridge in the form of an undulating tower rising directly out of the water and reaching a height equivalent to the width of the river. The second is a web of pedestrian walkways on two levels connecting a number of points on the river bank.

The vertical bridge is a slender column which is related to London as a city of spires with their horizons as the River Thames. This tower represents a 24-hour beacon of life and activity, visible as an orienting point from all over London. It is sited near the south bank in order to shift the virtual centre line of the river towards that side of the river.

The pedestrian web is an intricate lightweight urban structure dematerialising the horizontal. There are diverse and unexpected pathways within it forming a public membrane. The structure is a combination of pylons and tensile members creating a mutually supportive tensegrity structure over the river. The web contains pavilions, cafés, recreational places for winter and summer (covered and open) – a field of diverse spaces and events delicately suspended over the water.

The web generates a sense of adventure and excitement, celebrating the water and the sense of openness without obliterating or obscuring the drama of the Thames.

Architect   Daniel Libeskind
with   Robert Slinger, James Goodspeed,
                     Francis Henderson, Johannes Hucke,
                     Matthew Johnson, Santeri Lipasti,
                     Thomas Schropfer

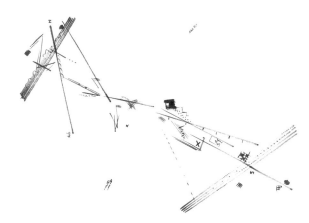

Fig. 257, *Preliminary drawing.* © Collection Libeskind

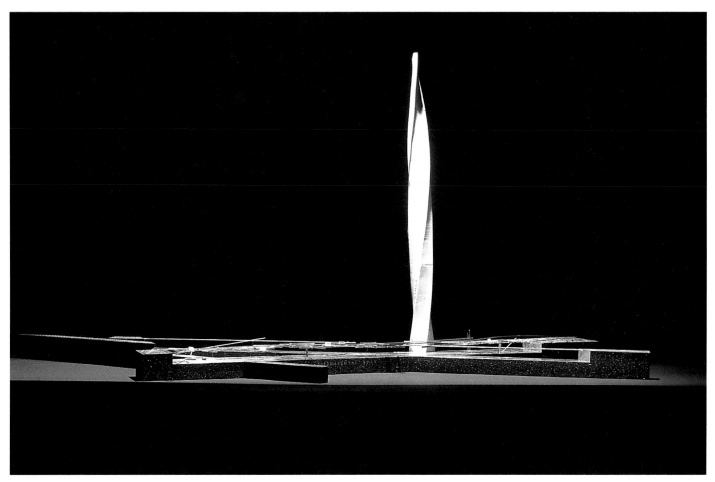

Fig. 256, *Model.* © Collection Libeskind

Fig. 258, *View from Waterloo Bridge: montage*. © Collection Libeskind

Fig. 259, *Drawing*. © Collection Libeskind

Fig. 260, *Drawing*. © Collection Libeskind

Fig. 261, *Drawing*. © Collection Libeskind

IAN RITCHIE ARCHITECTS

The bridge's main promenades are along both edges of the bridge, and are partly covered. Whichever the wind direction, a protected walk is always available. There are three cross-walks enabling people to move from one side to the other. These are aligned to create framed views of key London landmarks – St Paul's Cathedral, Big Ben and the Lloyd's building.

Covering the top of the bridge is the central garden, nearly one hectar in area, 30 metres wide and 277 metres long. It is laid out as a lawn, divided along its length by lines of water, one metre wide, which signify the primary trusses of the bridge. Small bridges allow the visitor to move easily from one zone to the other across the water strips. The three-dimensional nature of this landscape is created gardens, some grassed, others paved or gravelled which are sunken to provide protection from the wind.

Internally the spaces are on two levels, zoned by the main structural trusses into three ten metres wide bays running the length of the bridge and two perimeter bays each five metres wide. The spaces can accept passive recreation such as cinemas, restaurants and cafés with terraces and the occasional shop, and active recreations such as a family centre incorporating a swimming pool, ten-pin bowling, a gymnasium and a pool-side café. These proposals are indicative only of the bridge's potential to accommodate a diverse range of leisure facilities within single – and double – height spaces.

The structure is composed of steel plate trusses 277.5 metres long, which span the 160 metre wide navigational channel required by the Port of London Authority. At 9 metres in height they are deep enough to accommodate and support two generous floor levels within.

Architects	Ian Ritchie Architects: Simon Conolly, Christopher Hill, Helgo von Meier, Clarissa Matthews, Kuros Sarshar, Ian Ritchie, Robert Thum
Structural Engineers	Ove Arup & Partners John Thornton
Quality Surveyors and Cost Consultants	Hanscomb Mike Staples

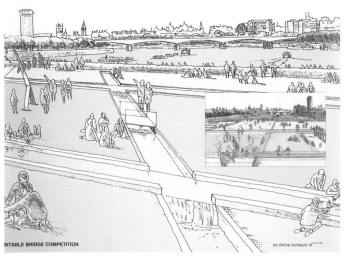

Fig. 262, *Perspective drawing*. © Collection Ian Ritchie Architects

Fig. 263, *Model*. © Collection Ian Ritchie Architects

Fig. 264, *Model*. © Collection Ian Ritchie Architects

152

Fig. 265,
*View from Waterloo
Bridge: montage.*
© Collection
Ian Ritchie Architects

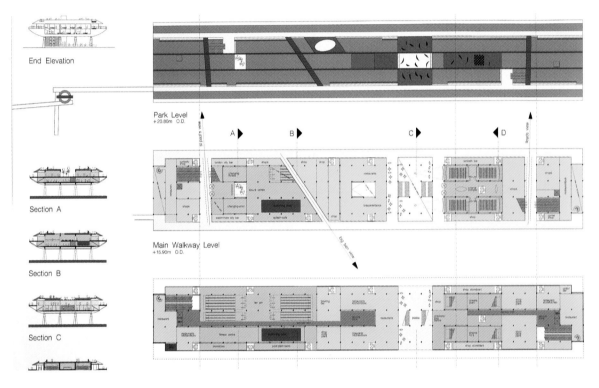

End Elevation

Section A

Section B

Section C

Park Level
+20.80m O.D.

A   B   C   D

Main Walkway Level
+15.90m O.D.

Fig. 266,
Plans.
© Collection
Ian Ritchie Architects

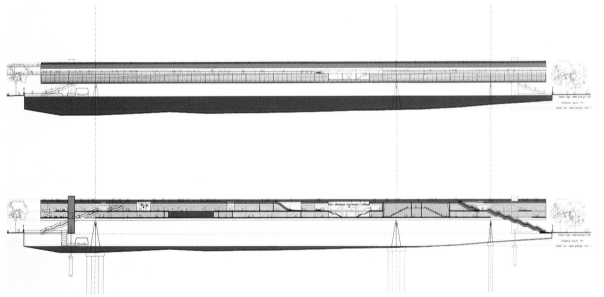

Fig. 267,
Longitudinal cross-section.
© Collection
Ian Ritchie Architects

153

Bibliography

General References

There are few publications devoted specifically to the inhabited bridge as a building type. The most comprehensive to date is DETHIER, Jean and EATON, Ruth, 'Inhabited Bridges', *Rassegna*, n°48 (special number), Bologna-Milan, 1991 (also in Italian and French eds.). See also, BOUGENNEC, Remy, *Ponts et merveilles: ponts habités en Europe*, Landerneau, 1992.

Thematic References

DETHIER, Jean, 'Past and present of inhabited bridges', *Rassegna*, n° 48, 1991, pp. 10-19

JACQUES, Annie, 'Triumphal bridges', *Rassegna*, n°48, 1991, pp. 51-53

MESQUI, Jean, 'The city and the bridge in Medieval Europe', *Rassegna*, n°48, 1991, pp. 20-25

References by City

AMIENS
KRIER, Rob, *The Reconstruction of the Historic Centre in Amiens and the Design of a new Inhabited Bridge*, 'Archives d'Architecture Moderne', Brussels, 1987

BATH
MANCO, Jean, 'Pulteney Bridge', *Architectural History*, vol 38, 1995, pp. 129-145

CHENONCEAUX
'Chenonceaux' *Connaissance des Arts*, n°37, (special number), 1995

DUBLIN
RICHARDSON, Margaret, 'The Dublin Gallery on the River Liffey by Luytens', *Rassegna*, n° 48, 1991, p. 83

FLORENCE
ROMBY, Giuseppina Carla, 'Ponte Vecchio in Florence', *Rassegna*, n° 48, 1991, pp. 59-61

LAUSANNE
Concours d'idées pour l'aménagement de quatre ponts habités à Lausanne, Lausanne, 1990

LONDON
The Story of three Bridges: The History of London Bridge, London, 1973

EATON, Ruth, 'Heads above water: the revival of inhabited bridges', *Perspectives on Architecture*, London, June 1995

HOME, Gordon, *Old London Bridge*, London, 1931

PRADEL, Jean-Louis, 'Le futur des anciens ponts habités: Londres relance avec faste une tradition oubliée', *L'Evènement du Jeudi*, Paris, 6-12 June 1996

LYON
GARDES, Gilbert, 'From symbolic bridge to bridge-garage', *Rassegna*, n° 48, 1991, pp. 81-82

NEW YORK
WILLIS, Carol, 'Inhabited Bridges for New York', *Rassegna*, n° 48, 1991, pp. 83-84

NOISIEL
MARREY, Bernard, 'Le Pont-moulin Meunier à Noisiel', *Rassegna*, n° 48, 1991, p. 65

PARIS
LEMOINE, Bertrand, 'Gustave Eiffel, a Bridge for the Exposition Universelle of 1878', *Rassegna*, n° 48, 1991, p. 82

MISLIN, Miron, 'Die überbauten Brücken von Paris: ihre bau- und stadtbaugeschichtliche Entwicklung im 12.-19. Jahrhundert', Ph.D Thesis, University of Stuttgart-Technische Hochschule, 1978

MISLIN, Miron, 'Paris, Île de la Cité: Die überbauten Brücken, *Storia della Città*, Milan, 1980

MISLIN, Miron, *Die überbauten Brücken: Pont Notre Dame; Baugestalt und Sozialstruktur*, Frankfurt, 1982

POTIÉ, Philippe, 'Du Cerceau project for the Pont Neuf in Paris: from the bridge street to the bridge square', *Rassegna*, n° 48, 1991, pp. 77-79

AMSTERDAM
BAETEN, Jean-Paul, 'Galman's project for a bridge on the Ij in Amsterdam', *Rassegna*, n° 48, 1991, pp. 81-82

SAN FRANCISCO
WILLIS, Carol, 'Mullgardt and San Francisco', *Rassegna*, n° 48, 1991, pp. 84-85

VENICE
BISA, Marco and MASOBELLO, Remigio, *Il ponte di Rialto: un restauro a Venezia*, Venice, 1991

CALABI, Donatella and MORACHIELLO, Paolo, *Rialto: le fabbriche e il ponte*, Turin, 1987

MORACHIELLO, Paolo, 'Venice Rialto Bridge', *Rassegna*, n° 48, 1991, pp. 71-73

TREIBER, Daniel, 'Louis Kahn's Palazzo dei Congressi in Venice', *Rassegna*, n° 48, 1991, p. 86

WOODWARD, Antony, 'Structures, rituals and romantic visions: bridges and eighteenth-century Venice', *Apollo*, September 1994, pp. 52-57

Photographic Credits

Alinari-Giraudon, figs. 18, 82
Archives du Loiret. All rights reserved. fig 41
Archives Municipales de Lyon, figs. 122, 126, 127, photos by St Jean, figs. 125, 142
© Bath Central Library, photo Fotek, fig. 93
© Bath Preservation Trust, fig. 95
Baugeschichtliches Archiv der Stadt Zurich, photos by A. Scherer, Fotograf, figs. 8, 40, 128
Beit Collection, Russborough, fig. 79
Berlin Stadtmuseum, Berlin photos by Hans-Joachim Bartsch, figs. 14-16
© Bernard Tschumi Architects, cover (verso), photo by Dan Cornish, fig. 29
Bibliothèque des Arts Decoratifs, Paris, Collection Maciet photos by Jean-Claude Planchet, figs. 12, 17, 19, 24, 37, 38, 75, 102, 103
Bibliothèque Municipale de Lyon, figs. 123, 124
Bibliothèque Nationale de France, Paris, figs. 33, 57, 64, 65
© Birmingham Museums and Art Gallery, figs. 23, 43
© British Library, figs. 128, 141
© British Museum, fig. 84
Château de Fere, fig. 44
© Nigel Coates, fig. 1
The Conway Library, Courtauld Institute of Art, fig. 100
Conservateur du Château de Chenonceau, figs. 45, 46
Ecole nationale supérieure des Beaux-Arts, Paris, figs. 105-107
© English Heritage, fig. 51
Ezra Stoller © Esto. All rights reserved, fig. 11
Gemeentelijke Archiefdienst Amsterdam, fig. 132
Giraudon, Paris, fig. 85
David Grandorge, photos by, figs. 6, 7
© Guildhall Library, Corporation of London, cover (recto), photos by Geremy Butler, figs. 21, 48, 136, 145-146
© Her Majesty The Queen, fig. 89
© Jörg P. Anders, Berlin, fig. 68
MNAM-CCI, Centre Georges Pompidou, Paris photos by Jean-Claude Planchet, figs. 13, 143
© The Metropolitan Museum of Art, New York, fig. 80
Musée Carnavalet, figs. 27, 56, 58-60, 63, 66, 67, 69
© Museen der Stadt Wien, fig. 109
© Museum of London, figs. 2, 49, 50, 52, 53, 54
Musée Nissim de Camondo, photo by Hugo Maertens, fig. 74
© National Gallery of Canada, Ottawa, fig. 22
Newcastle City Libraries and Arts, figs. 9, 31, 32
New York Public Library, figs. 139, 140
© Photothèque des Musées de la Ville de Paris, cover (recto), figs. 61, 62, 73, 104
Prudence Cuming Ass. Ltd., figs. 47, 55, 98
© Rémy Bouguennec 1996 Ponts et merveilles Carré noir F-29800 Landerneau, fig. 34
© Royal Institute of British Architects, photos by A. C. Cooper, figs. 88, 90, 110, 111, 120, 121, 131, 135

© RMN – Arnaudet, Paris, fig. 91
© Sir John Soane's Museum photos by Geremy Butler, figs. 94, 99, 112-118
Staatliche Kunsthalle Karlsruhe, fig. 76
Staatliche Museen zu Berlin, Kupferstichkabinett, fig. 77
© Staatsarchiv Basel-Stadt, fig. 134a
© Stadtarchiv Esslingen am Neckar, photo by Heinzmann, fig. 36
© Tate Gallery, London, figs. 25, 119
© Victoria Art Gallery, Bath and North East Somerset Council photos by Fotek, figs. 95, 96

Post 1945

© Allies & Morrison, fig. 195, 196 photo by Michael Dyer Ltd
© Alsop & Störmer, figs. 215-218, Principle in charge, Prof. W. Alsop, Project Architect, G. Tsoutsos, Drawing, Imke Woelk, Computer Image, Steve Bedford
© Cezary Bednarski, figs. 201-204
© Mario Bellini Associati, figs. 205, 206
© Edward Cullinan Associates, fig. 193
© Terry Farrell & Partners, figs. 212-214, photo by Nigel Young
© Yona Friedman, figs. 152, 153
© Gunther Feuerstein, figs. 154-156
David Grandorge, figs. 222, 223, 233, 241, 246, 256, 263, 264
© Michael Graves, figs. 157, 159, photo by Acme photo, fig. 158, photo by Ted Bichford
Simon Hazelgrove, fig. 221
© Richard Horden Associates, figs. 197-200, photo by A. O'Mahoney
© Hunt Thompson Associates, fig. 192
Image Point, London, photographed figs. 224-232, 234-236, 242-244, 247-255, 258-262, 265-267
© Jellicoe and Coleridge, fig. 147, photo by A. C. Cooper, figs. 148-151
© Rob Krier, figs. 164-169
© Levitt Bernstein Associates, fig. 191, photo by Michael Dyer Ltd
© Marks and Barfield, figs. 177-179
© Morphosis, figs. 188-190
© John Outram Associates, fig. 194, photo by Michael Dyer Ltd
© Cedric Price, fig. 170
© Wellington Reiter, figs. 207-211
© Richard Rogers Parnership, figs. 171-176
© R. Seifert & Partners, figs. 160-163
© SITE environmental Design, figs. 180, 181, photos by Studio Natori, figs. 182-184, photos by Kazuo Natori
© Team Luscher Switzerland, figs. 185-187
A & B Photographic Services Ltd. London kindly assisted.

Friends of the Royal Academy

SPONSORS

Mrs Denise Adeane
Mr M.R. Anderson
Mr and Mrs Theodore Angelopoulos
Mr P.F.J. Bennett
Mrs D. Berger
Mr David Berman
Mr and Mrs George Bloch
Mrs J. Brice
Mr Jeremy Brown
Mrs Susan Burns
Mr and Mrs P.H.G. Cadbury
Mrs L. Cantor
Mrs Denise Cohen
Mrs Elizabeth Corob
Mr and Mrs S. Fein
Mr M. J. Fitzgerald
    (Occidental International Oil Inc.)
Mr and Mrs R. Gapper
Mr and Mrs Robert Gavron
Mr and Mrs Michael Godbee
Lady Gosling
Lady Grant
Mr Harold Joels
Mrs G. Jungels-Winkler
Mr J. Kirkman
Dr Abraham Marcus
Mrs Xanna De Mico
The Oakmoor Trust
Ocean Group p.l.c. (P.H. Holt Trust)
Mr and Mrs Godfrey Pilkington
Mr and Mrs G.A. Pitt-Rivers
The Worshipful Company of Saddlers
Mr and Mrs David Shalit
Mrs Roama Spears
The Stanley Foundation
Mr Helmut Sternberg
Mrs Paula Swift
Mr Robin Symes
Mrs Edna S. Weiss
Mrs Linda M. Williams
Sir Brian Wolfson

ASSOCIATE SPONSORS

Mr Richard B. Allan
Mr Richard Alston
Mr Ian F.C. Anstruther
Mrs Ann Appelbe
Mr John R. Asprey
Lady Attenborough
Mr J.M. Bartos
Mrs Susan Besser
Mrs Linda Blackstone
Mrs C.W.T. Blackwell
Mr Peter Boizot
C.T. Bowring (Charities Trust) Ltd
Mrs J.M. Bracegirdle
Mr Cornelius Broere
Lady Brown
Mr P.J. Brown Jr
Mr T.M. Bullman
Mr and Mrs James Burt
Mrs A. Cadbury
Mr and Mrs R. Cadbury
Mrs C.A. Cain
Miss E.M. Cassin
Mr R.A. Cernis
Mr. S. Chapman
Mr W.J. Chapman
Mrs J.V. Clarke
Mr John Cleese
Mrs R. Cohen
Mrs N.S. Conrad
Mr and Mrs David Cooke
Mr C. Cotton
Mrs Saeda H. Dalloul
Mr and Mrs D. de Laszlo
Mr John Denham
The Marquess of Douro
Mr D.P. Duncan
Mr Kenneth Edwards
Mrs K.W. Feesey MSc
Dr G.-R. Flick
Mr J.G. Fogel

Mr Graham Gauld
Mr Stephen A. Geiger
Mrs R.H. Goddard
Mrs P. Goldsmith
Mr Gavin Graham
Mr and Mrs R.W. Gregson-Brown
Mrs O. Grogan
Mr J.A. Hadjipateras
Mr B.R.H. Hall
Mr and Mrs Richard Harris
Miss Julia Hazandras
Mr Malcolm Herring
Mrs P. Heseltine
Mrs K.S. Hill
Mr R.J. Hoare
Mr Reginald Hoe
Mr Charles Howard
Mrs A. Howitt
Mr Norman J. Hyams
Mr David Hyman
Mrs Manya Igel
Mr C.J. Ingram
Mr S. Isern-Feliu
The Rt. Hon. The Countess of Iveagh
Mrs I. Jackson
Lady Jacobs
Mr and Mrs S.D. Kahan
Mr and Mrs J. Kessler
Mr D.H. Killick
Mr P.W. Kininmonth
Mrs L. Kosta
Mrs E. Landau
Mr and Mrs M.J. Langer
Mrs J.H. Lavender
Mr and Mrs Andrew D. Law
Mr Morris Leigh
Mr J.R.A. Leighton
Mr Owen Luder
Mrs G.M.S. McIntosh
Mr Peter I. McMean
Mrs Susan Maddocks
Ms R. Marek
The Hon. Simon Marks
Mr and Mrs V.J. Marmion
Mr B.P. Marsh
Mr and Mrs J.B.H. Martin
Mr R.C. Martin
Mr and Mrs G. Mathieson
Mr J. Menasakanian
Mr J. Moores
Mrs A. Morgan
Mr A.H.J. Muir
Mr David H. Nelson
Mrs E.M. Oppenheim-Sandelson
Mr Brian R. Oury
Mrs J. Palmer
Mrs J. Pappworth
Mr J.H. Pattisson
Mrs M.C.S. Philip
Mrs Anne Phillips
Mr Ralph Picken
Mr G.B. Pincus
Mr W. Plapinger
Mrs J. Rich
Mr Clive and Mrs Sylvia Richards
Mr F.P. Robinson
Mr M. Robinson
Mr D. Rocklin
Mrs A. Rodman
Lady Rootes
Mr and Mrs O. Roux
The Hon. Sir Stephen Runciman CH
Sir Robert Sainsbury
Mr G. Salmanowitz
Mr Anthony Salz
Lady Samuel
Mrs Bernard L. Schwartz
Mr Mark Shelmerdine
Mrs Emma Shulman
Mr R.J. Simmons
Mr John H.M. Sims
Dr and Mrs M.L. Slotover
The Spencer Wills Trust
Mr and Mrs J.G. Studholme
Mr J.A. Tackaberry
Mr N. Tarling
Mr G.C.A. Thom
Mrs Andrew Trollope
Mr A.J. Vines
Mrs C.H. Walton

Mr D.R. Walton Masters
Mr Neil Warren
Miss J. Waterous
Mrs Roger Waters
Mrs J.M. Weingarten
Mrs C. Weldon
Mr Frank S. Wenstrom
Mr Julyan Wickham
Mrs I. Wolstenholme
Mr W.M. Wood
Mr R.M. Woodhouse
Mr and Mrs F.S. Worms

Royal Academy Trust

BENEFACTORS

H.M. The Queen
Mr and Mrs Russell B. Aitken
American Airlines
The Annie Laurie Aitken Charitable Trust
American Associates of the Royal Academy Trust
American Express Company
Mrs John W. Anderson II
The Andor Family
The Hon. and Mrs Walter H. Annenberg
Mr Walter Archibald
Marilyn B. Arison
The Hon. Anne and Mr Tobin Armstrong
Asprey
AT & T
AT & T (UK) Ltd
Barclays Bank plc
Mr and Mrs Sid R. Bass
Mr Tom Bendhem
Benihana Group
Mrs Brenda Benwell-Lejeune
Mr David Berman
In Memoriam: Ida Rose Biggs
Charlotte Bonham-Carter Charitable Trust
Denise and Francis Booth
British Airways, North America
British Gas plc
The British Petroleum Company plc
BP America
British Steel plc
Mr Keith Bromley
The Brown Foundation Inc.BT
BUNZL plc
Iris and B. Gerald Cantor
Sir Richard Carew Pole
The Rt. Hon. the Lord Carrington
The Trustees of the Clore Foundation
The Cohen Family Charitable Trust
The John S. Cohen Foundation
The Ernest Cook Trust
Mrs John A. Cook
Crabtree & Evelyn
The Hon. and Mrs C. Douglas Dillon
Sir Harry and Lady Djanogly
In Memoriam: Miss W.A. Donner
The Dulverton Trust
Alfred Dunhill Limited
Miss Jayne Edwardes
The John Ellerman Foundation
Mr E.A. Emerson
English Heritage
The Eranda Foundation
The Esmée Fairbairn Charitable Trust
Esso UK PLC
Lord and Lady Faringdon
Mr and Mrs Eugene V. Fife
Mr and Mrs Donald R. Findlay
Mr Walter Fitch III
Mrs Henry Ford II
The Henry Ford II Fund
The Foundation for Sport and the Arts
The Late John Frye Bourne
The Garfield Weston Foundation
Gartmore plc
The Gatsby Foundation
The Getty Grant Program
The J. Paul Getty Jr Trust
The Lady Gibson
Glaxo Wellcome plc
The Jack Goldhill Charitable Trust

Maurice and Laurence Goldman
The Horace W. Goldsmith Foundation
The Worshipful Company of Goldsmiths
The Greentree Foundation
The Worshipful Company of Grocers
The Worshipful Company of Haberdashers
The Paul Hamlyn Foundation
The Late Dr and Mrs Armand Hammer
Mrs Sue Hammerson
Philip and Pauline Harris Charitable Trust
Mr and Mrs Gustave Hauser
The Hayward Foundation
Mr and Mrs Randolph Hearst
Klaus and Belinda Hebben
The Hedley Foundation
Mrs Henry J. Heinz II The Henry J. and Drue
    Heinz Foundation
Drue Heinz Trust
The Heritage of London Trust
The Howser Foundation
The Idlewild Trust The J.P. Jacobs Charitable Trust
Jerwood Foundation
Mr and Mrs Donald P. Kahn
The Kresge Foundation
The Kress Foundation
Mr and Mrs Sol Kroll
Ladbroke Group Plc
Mr D.E. Laing
The Kirby Laing Foundation
The Maurice Laing Foundation
The Landmark Hotel
The Landmark Trust
The Lankelly Foundation
Mr John S. Latsis
The Leche Trust
The Leverhulme Trust Mr Leon Levy and
    Ms Shelby White
Lex Service Plc
The Linbury Trust
The Ruth and Stuart Lipton Charitable Trust
Sir Sydney and Lady Lipworth
Mr John Madejski
Mrs T.S. Mallinson
The Manifold Trust
The Stella and Alexander Margulies
Charitabl Trust
Mr and Mrs John L. Marion
Marks & Spencer
Mrs Jack C. Massey
M.J. Meehan & Company
Mr. Paul Mellon KBE
The Anthony and Elizabeth Mellows
    Charitable Trust
The Mercers' Company
Mr and Mrs Donald Moore
The Henry Moore Foundation
Museums and Galleries Improvement Fund
National Westminster Bank PLC
Diane A. Nixon The Normanby Charitable Trust
Otemae College
    The Peacock Charitable Trust
Mr and Mrs Frank Pearl
The Pennycress Trust
In Memoriam: Mrs Olive Petit
The P.F. Charitable Trust
The Pilgrim Trust
Mr A.N. Polhill
The Hon. and
Mrs Leon B. Polsky
Provident Financial plc
The Radcliffe Trust
The Rayne Foundation
Mr and Mrs Laurance S. Rockefeller
The Ronson Charitable Foundation
Mr and Mrs Leonard Rosoman
Rothmans UK Holdings Limited
The J. Rothschild Group Charitable Trust
Rothschilds Inc
Royal Mail International
The RTZ - CRA Group The Late
    Dr Arthur M. Sackler
Mrs Arthur M. Sackler
The Sainsbury Family Charitable Trusts
Mrs Jean Sainsbury
Mrs Basil Samuel
Save & Prosper Educational Trust
Mrs Frances G. Scaife
Sea Containers Limited
Sheeran Lock
Shell UK Ltd

The Archie Sherman Charitable Trust
Mr and Mrs James C. Slaughter
The Late Mr Robert Slaughter
Pauline Denyer Smith and Paul Smith CBE
Sotheby's The Spencer Charitable Trust
Miss K. Stalnaker
The Starr Foundation
The Steel Charitable Trust
Bernard Sunley Charitable Foundation
Lady Judith Swire
Mr and Mrs A. Alfred Taubman
Mr and Mrs Vernon Taylor Jr.
Texaco Inc
Time Out magazine
G. Ware and Edythe Travelstead
Seiji Tsutsumi
    The Douglas Turner Charitable Trust
    The 29th May 1961 Charitable Trust
    Unilever PLC
The Weldon UK Charitable Trust
Mr and Mrs Keith S. Wellin
The Welton Foundation
Westminster City Council
Mr and Mrs Garry H. Weston
The Hon. and Mrs John C. Whitehead
Mrs John Hay Whitney
Mr Frederick B. Whittemore
Mr and Mrs Wallace S. Wilson
The Wolfson Foundation
The Late Mr Charles Wollaston
The Late Mr Ian Woodner
Mr and Mrs William Wood Prince

Corporate Membership Scheme

CORPORATE PATRON

Glaxo Wellcome plc

CORPORATE MEMBERS

Alliance & Leicester Building Society
All Nippon Airways Co. Ltd.
Arthur Andersen & Andersen Consulting
Ashurst Morris Crisp
Atlantic Plastics Limited
A.T. Kearney Limited
Bankers Trust
Bank Julius Baer & Co Ltd
Banque Indosuez
Barclays Bank plc
BAT Industries plc
BMW (GB) Limited
BP Chemicals
British Aerospace PLC
British Airways
British Alcan Aluminium plc
British Gas plc
Bunzl plc
BUPA
Cantor Fitzgerald
Christie's
Chubb Insurance Company
Cookson Group plc
Coopers & Lybrand
Courage Limited
C.S. First Boston Group
The Daily Telegraph plc
Datastream International
Department of National Heritage
The Diamond Trading Company
Dow Jones Telerate Ltd
Eaga Ltd
Robert Fleming & Co Limited
Gartmore Investment Management plc
Goldman Sachs International Limited
Grand Metropolitan plc
Guinness PLC
Hay Management
Consultants Limited
Hillier Parker May &
Rowden
IBM
ICI

Industrial Bank of Japan, Limited
Jaguar Cars Ltd
Kvaerner Construction Ltd
John Laing plc
Lehman Brothers International
Lloyds Private Banking Limited
E.D. & F. Man Limited
    Charitable Trust
M & G Group P.L.C.
Marks & Spencer
Merrill Lynch Europe Ltd
Midland Bank
MoMart plc
Morgan Guaranty Trust Company, New York
Morgan Stanley International
Pearson plc
The Peninsular and Oriental Steam Navigation Co
Pentland Group plc
The Reader's Digest Association
Republic National Bank of New York
Reuters
Rothmans UK Holdings Limited
The Royal Bank of Scotland plc
The RTZ-CRA Group
Salomon Brothers
Santa Fe Exploration (U.K.) Limited
Sea Containers Ltd.
Silhouette Eyewear
SmithKline Beecham
The Smith & Williamson Group
Société Générale, UK
Southern Water plc
TI Group plc
Unilever UK Limited

CORPORATE ASSOCIATES

ABL Group
AT & T
Bass PLC
BHP Petroleum Ltd
BMP DDB Needham
The BOC Group
Booker plc
Bovis Construction Limited
Charterhouse plc
CJA (Management Recruitment
    Consultants) Limited
Clifford Chance
Coutts & Co
Credit Lyonnais Laing
The Dai-Ichi Kangyo Bank, Ltd
Dalgleish & Co
De La Rue plc
Durrington Corporation Limited
Enterprise Oil plc
Fina plc
Foreign & Colonial Management Ltd
General Accident plc
The General Electric Company plc
Guardian Royal Exchange plc
H.J. Heinz Company Limited
John Lewis Partnership plc
Kleinwort Benson Charitable Trust
Lex Service PLC
Linklaters & Paines
Macfarlanes
Mars G.B. Limited
Nabarro Nathanson
NEC (UK) Ltd
Newton Investment Management Limited
Nortel Ltd
Ove Arup Partnership
The Rank Organisation Plc
Reliance National Insurance Company (UK) Ltd
Royal Insurance Holdings plc
Sainsbury's PLC
Schroders plc
J. Henry Schroder & Co Limited
Sears plc
Sedgwick Group plc
Slough Estates plc
Sotheby's
Sun Life Assurance Society plc
Tate & Lyle Plc
Tomkins PLC
Toyota Motor Corporation
United Biscuits (UK) Limited

Sponsors of Past Exhibitions

The Council of the Royal Academy thanks sponsors of past exhibitions for their support. Sponsors of major exhibitions during the last ten years have included the following:

Alitalia
Italian Art in the 20th Century
1989

Allied Trust Bank
Africa: The Art of a Continent
1995

American Express Foundation
Je suis le cahier: The Sketchbooks
of Picasso 1986

Anglo American Corporation of South Africa
Africa: The Art of a Continent 1995

The Banque Indosuez Group
Pissarro: The Impressionist and the City 1993

Banque Indosuez and W.I. Carr
Gauguin and The School of Pont-Aven:
Prints and Paintings 1989

BBC Radio One
The Pop Art Show 1991

BMW (GB) Limited
Georges Rouault: The Early Years,
1903-1920 1993
David Hockney: A Drawing Retrospective 1995

Bovis Construction Ltd
New Architecture 1986

British Airways
Africa: The Art of a Continent 1995

British Alcan Aluminium
Sir Alfred Gilbert 1986

British Petroleum Company plc
British Art in the 20th Century 1987

BT
Hokusai 1991

Canary Wharf Development
New Architecture 1986

Cantor Fitzgerald
From Manet to Gauguin:
Masterpieces from Swiss Private
Collections 1995

The Capital Group Companies
Drawings from the J. Paul Getty Museum 1993

The Chase Manhattan Bank
Cézanne: the Early Years 1988

Chilstone Garden Ornaments
The Palladian Revival: Lord Burlington
and his House and Garden at Chiswick 1995

Christie's
Frederic Leighton 1830-1896 1996

Classic FM
Goya: Truth and Fantasy,
The Small Paintings 1994
The Glory of Venice: Art in the
Eighteenth Century 1994

Corporation of London
Living Bridges 1996

The Dai-Ichi Kangyo Bank Limited
222nd Summer Exhibition 1990

The Daily Telegraph
American Art in the 20th Century 1993

De Beers
Africa: The Art of a Continent 1995

Deutsche Morgan Grenfell
Africa: The Art of a Continent 1995

Digital EquipmentCorporation
Monet in the '90s: The Series Paintings 1990

The Drue Heinz Trust
The Palladian Revival: Lord Burlington and
his House and Garden at Chiswick 1995

The Dupont Company
American Art in the 20th Century 1993

The Economist
Inigo Jones Architect 1989

Edwardian Hotels
The Edwardians and After: Paintings and
Sculpture from the Royal Academy's
Collection, 1900-1950 1990

Electricity Council
New Architecture 1986

Elf
Alfred Sisley 1992

Esso Petroleum Company Ltd
220th Summer Exhibition 1988

Fiat
Italian Art in the 20th Century 1989

Financial Times
Inigo Jones Architect 1989

Fondation Elf
Alfred Sisley 1992

Ford Motor Company Limited
The Fauve Landscape:
Matisse, Derain, Braque and their Circle 1991

Friends of the Royal Academy
Sir Alfred Gilbert 1986

Gamlestaden
Royal Treasures of Sweden, 1550-1700 1989

Joseph Gartner
New Architecture 1986

Générale des Eaux Group
Living Bridges 1996

J. Paul Getty Jr Charitable Trust
The Age of Chivalry 1987

Glaxo Wellcome plc
From Byzantium to El Greco 1987
Great Impressionist and other Master Paintings
from the Emil G. Bührle Collection, Zurich 1991

The Guardian
The Unknown Modigliani 1994

Guinness PLC
Twentieth-Century Modern Masters: The Jacques
and Natasha Gelman Collection 1990
223rd Summer Exhibition 1991
224th Summer Exhibition 1992
225th Summer Exhibition 1993
226th Summer Exhibition 1994
227th Summer Exhibition 1995
228th Summer Exhibition 1996

Guinness Peat Aviation
Alexander Calder 1992

Harpers & Queen
Georges Rouault: The Early Years, 1903-1920 1993
Sandra Blow 1994
David Hockney: A Drawing Retrospective 1995
Roger de Grey 1996

The Henry Moore Foundation
Henry Moore 1988

Alexander Calder 1992
Africa: The Art of a Continent 1995

The Independent
The Art of Photography 1839-1989 1989
The Pop Art Show 1991
Living Bridges 1996

Industrial Bank of Japan, Limited
Hokusai 1991

Intercraft Designs Limited
Inigo Jones Architect 1989

Joannou & Paraske-Vaides (Overseas) Ltd
From Byzantium to El Greco 1987

The Kleinwort Benson Group
Inigo Jones Architect 1989

Lloyds Bank
The Age of Chivalry 1987

Logica
The Art of Photography, 1839-1989 1989

The Mail on Sunday
Royal Academy Summer Season 1992
Royal Academy Summer Season 1993

Marks & Spencer
Royal Academy Schools Premiums 1994
Royal Academy SchoolsFinal Year Show 1994

Martini & Rossi Ltd
The Great Age of British Watercolours,
1750-1880 1993

Paul Mellon KBE
The Great Age of British Watercolours,
1750-1880 1993

Mercury Communications
The Pop Art Show 1991

Merrill Lynch
American Art in the 20th Century 1993

Midland Bank plc
The Art of Photography 1839-1989 1989
RA Outreach Programme 1992-1996
Lessons in Life 1994

Minorco
Africa: The Art of a Continent 1995

Mitsubishi Estate Company UK Limited
Sir Christopher Wren and the Making
of St Paul's 1991

Mobil
From Byzantium to El Greco 1987

Natwest Group
Reynolds 1986
Nicolas Poussin 1594-1665 1995

Olivetti
Andrea Mantegna 1992
Otis Elevators
New Architecture 1986

Park Tower Realty Corporation
Sir Christopher Wren and the Making
of St Paul's 1991

Pearson plc
Eduardo Paolozzi Underground 1986

Pilkington Glass
New Architecture 1986

Premiercare (National Westminster Insurance Services)
Roger de Grey 1996

Redab (UK) Ltd
Wisdom and Compassion:
The Sacred Art of Tibet 1992

Reed International plc
Toulouse-Lautrec: The Graphic Works 1988
Sir Christopher Wren and the Making
    of St Paul's 1991

Republic National Bank of New York
Sickert: Paintings 1992

The Royal Bank of Scotland
The Royal Academy Schools Final Year
    Show 1996

Arthur M. Sackler Foundation
Jewels of the Ancients 1987

Salomon Brothers
Henry Moore 1988

The Sara Lee Foundation
Odilon Redon: Dreams and Visions 1995

Sea Containers Ltd
The Glory of Venice:
    Art in the Eighteenth Century 1994

Silhouette Eyewear
Egon Schiele and His Contemporaries:
    From the Leopold Collection, Vienna 1990
Wisdom and Compassion:
    The Sacred Art of Tibet 1992
Sandra Blow 1994
Africa: The Art of a Continent 1995
Société Générale, UK
Gustave Caillebotte: The Unknown Impressionist 1996

Société Générale de Belgique
Impressionism to Symbolism: The Belgian
    Avant-Garde 1880-1900 1994

Spero Communications
The Royal Academy Schools Final Year Show 1992

Texaco
Selections from the Royal Academy's Private
    Collection 1991

Thames Water Plc
Thames Water Habitable Bridge Competition 1996

The Times
Old Master Paintings from the
    Thyssen-Bornemisza Collection 1988
Wisdom and Compassion:
    The Sacred Art of Tibet 1992
Drawings from the J. Paul Getty Museum 1993
Goya: Truth and Fantasy,
    The Small Paintings 1994
Africa: The Art of a Continent 1995

Tractabel
Impressionism to Symbolism:
    The Belgian Avant-Garde 1880-1900 1994

Unilever
Frans Hals 1990

Union Minière
Impressionism to Symbolism:
    The Belgian Avant-Garde 1880-1900 1994

Vistech International Ltd
Wisdom and Compassion: The Sacred Art
    of Tibet 1992

Other Sponsors

Sponsors of events, publications and other items
in the past two years:

Academy Group Limited
Agnew's
Air Hong Kong
Air Jamaica
Air UK
Alitalia
Allied Trust Bank
Arthur Andersen
John A. Anderson
Athenaeum Hotel and Apartments
Austrian Airlines
Mr and Mrs Martin Beisly
The Beit Trust
Berggruen & Zevi Limited
The Britto Foundation
The Brown Foundation
Bulgari Jewellery
James Butler RA
Cable & Wireless
The Calouste Gulbenkian Foundation (Lisbon)
Cathay Pacific
Chilstone Garden Ornaments
Christopher Wood Gallery
Citibank N.A.
Terance Cole
Columbus Communications
Condé Nast Publications
Mrs Shimona Cowan
Deutsche Morgan Grenfell
Hamish Dewar
Jennifer Dickson RA
Sir Harry and Lady Djanogly
D.W. Viewboxes Ltd
The Elephant Trust
Brenda Evans
Sebastian de Ferranti
Fina Plc
FORBES Magazine, New York
Forte Plc
The Four Seasons Hotels
Isabel Goldsmith
Ivor Gordon
Lady Gosling
Julian Hartnoll
Ken Howard RA
IBM UK Limited
Inter-Continental Hotels
Intercraft Designs Limited
Jaguar Cars Limited
John Lewis Partnership plc
A.T. Kearney Limited
KLM
Count and Countess Labia
The Leading Hotels of the World
The Leger Galleries, London
The A.G. Leventis Foundation
Mr and Mrs J.H.J. Lewis
The Maas Gallery
Mandarin Oriental Hotel Group
Martini & Rossi Ltd
Masterpiece
Mercury Communications Ltd
Merrill Lynch
Micheal Hue-Williams Fine Art Limited
NK
The Nigerian Friends of africa95

Novell U.K. Ltd
Richard Ormond
Patagonia
Penshurst Press Ltd
Mr and Mrs James Phelps
Stuart Pivar
Polaroid (UK) Ltd
Price Waterhouse
The Private Bank & Trust Company Limited
Ralph Lauren
The Regent Hotel
The Robina Group
The Rockefeller Foundation
N. Roditi & Co.
Royal Mail International
Mrs Basil Sauel
Sears plc
Simon Dickinson Ltd
Peyton Skipwith
Swan Hellenic Ltd
Mr and Ms Daniel Unger
Kurt Unger
Vista Bay Club Seychelles
Vorwerk Carpets Limited
John Ward RA
Warner Bros.
W S Yeates plc
Mrs George Zakhem
ZFL